CABLE TELEVISION
IN THE CITIES

CABLE TELEVISION IN THE CITIES:

community control, public access, and minority ownership

charles tate, editor

THE URBAN INSTITUTE • WASHINGTON, D.C.

The Urban Institute is a nonprofit research organization established in 1968 to study problems of the nation's urban communities. Independent and nonpartisan, the Institute responds to current needs for disinterested analyses and basic information and attempts to facilitate the application of this knowledge. As part of this effort, it cooperates with federal agencies, states, cities, associations of public officials, the academic community and other sectors of the general public.

The Institute's research findings and a broad range of interpretive viewpoints are published as an educational service. Conclusions expressed in Institute papers are those of the authors and do not necessarily reflect the views of other staff members, officers or directors of the Institute, or of organizations which provide funds toward support of Institute studies.

LC 70-184624
ISBN 87766-020-4
UI 91-201-1

First Printing, December 1971
Second Printing, March 1972

List Price: $3.95

The Urban Institute
2100 M Street, N.W.
Washington, D.C. 20037

CONTENTS

CONTENTS

SECTION 1

ACKNOWLEDGMENTS

Many people have made contributions to this project. The following deserve special recognition: Nathaniel Lacy, Doris Ellerbe, Ray Lawrence, Morrissa Young, Connie Frye, Olivia Reaves, Robert Artisst, Valerie Mobley, Tina Williams, Gil Mendleson, Valerie Bullock, and Elaine Worsley.

Tony Perot, former director of the Urban Fellows Program at The Urban Institute is responsible for initiating the work that stimulated the interest and activity that led to this publication. The early involvement and support of Berkeley Burrell and his staff at the National Business League has been invaluable. William Gorham, President of The Urban Institute, has provided consistent encouragement and support.

research and documentation staff

Charles Tate, director

Fran Rusan, associate director

consultants

Mimi Hayes, general editing

Adrienne Manns, writer

Tina McIntosh, cataloging

Eric Stark, research

support

Brenda Brown

Diana Dudley

Barbara Hines

PREFACE

Public and private resources committed to the economic revitalization and development of urban ghettoes in America are inadequate. Neither the War on Poverty nor the current programs to create 'model cities' or promote 'black capitalism' have provided the comprehensive development program and the resources required to eliminate the chronic poverty that exists.

Repeated appeals have been made for a Marshall Plan for the Ghettoes, but no proposals or plans on this scale have been introduced by the present administration or Congress. Current political and economic trends drastically reduce the possibilities for a change in this situation in the foreseeable future.

Because of inadequate funding, community development corporations and organizations are forced to carry out palliative, pacification projects. The long-term development programs needed to combat deteriorating physical, economic and social conditions in Harlem, Hough, Watts, Detroit and other urban ghettoes cannot be implemented with the resources provided.

'Ghetto dispersal' has been advanced by some social theorists as a less costly solution to the problem of ghetto underdevelopment. Others advocate individual 'escape' for ghetto residents through education, training, and higher paying jobs in white corporations. Neither of these alternatives will provide a just and equitable solution for the growing minority populations in the cities. Neither addresses racism and the concentration of wealth and power in America as underlying causal factors for ghetto underdevelopment.

The struggle of minorities for self-determination and self-reliance has produced many positive benefits. There is a clearer understanding of the causes of impoverishment and powerlessness, stronger leadership in local communities, increased technical and management expertise, greater recognition of the need for cooperative activity, and a more sophisticated use of political power.

There have been very few opportunities, however, to consolidate these gains through a community-wide development program. The difficulty lies in finding durable development projects that will thrive and grow on indigenous resources, thus increasing technical skills and economic and political leverage.

The physical plant of the ghetto has been over-depreciated and worn-out. Hence, the range of viable economic activity is severely limited. Housing rehabilitation, service and retail ventures, and small manufacturing operations account for the bulk of the projects now being attempted. But the growth potential of these enterprises is low and the failure rate is high.

A number of emerging new service industries are better vehicles for community-wide development including day care, low-income housing management, health services, and communications. Each of these enterprises can potentially enjoy a community-wide market, utilize technical skills of local residents with short-term training, and employ traditional business management. One of the most potent of these new service opportunities is cable television.

Most often, minority development organizations are unaware of or unable to acquire sufficient information and data about these new opportunities in time to design and plan a development approach before affluent entrepreneurs corner the market. This reference and guide has been prepared to overcome that obstacle and to enhance the opportunities for minority groups in cable television.

Why cable television? A revolution in electronic communications systems is well underway. It is now possible to provide every village, hamlet, neighborhood, community, city, and town with a local, people-oriented television and radio system that is responsive to and reflective of the differences in culture, language, history, experience, and race. The means also exist to interconnect these systems nationally and internationally—to establish an effective communications link between people of all nations, races and cultures.

Audio and video messages are routinely beamed from the surface of the moon to millions of listeners and viewers around the world. Fifty-five nations signed an agreement on August 20, 1971 establishing a new, permanent International Telecommunications Satellite organization. The Federal Communications Commission recently ended the AT&T and Western Union monopoly on microwave relay data transmission and enabled several new companies to compete in this area. The result is likely to be the interconnection of computer facilities on a nationwide basis. Proposals are now being submitted for a domestic satellite system. A worldwide revolution in communications is taking place.

An integral component of this global explosion in communications is Community Antenna Television (CATV), also referred to as cable television or simply cable. It is this system that could enable America's minorities to challenge the communication systems that exploit the ghettoes, barrios, and reservations. Control, ownership and operation of cable systems by minorities could provide economic and political leverage, and the management and technical expertise required to accomplish a dramatic break in the cycle of dependency and exploitation.

Imagine television and radio systems where blacks could program for blacks; Chicanos for Chicanos; Indians for Indians and Puerto Ricans for Puerto Ricans—a system that can give the community a communications voice as well as the income and profit that the system receives for providing this service.

Right now, all of this and much more is technologically feasible and economically practical. Two features of CATV make this so. First, the enormous signal-carrying or channel capacity of the system. Second, the subscriber-oriented economic structure of the business.

The system can provide from 12 to 60 channels of interference-free signals, thus creating an abundance of channels instead of the 'scarcity' that characterizes the present system of commercial over-the-air television. Unlike commercial television, cable does not rely on advertising or advertisers for its revenues. It is a subscriber-supported system like the telephone company.

Because of the number of channels, the viewing public can dictate not only the style and type of programming it wants, but can actually produce and participate in that programming. "The medium can be the message."

The potential for nonentertainment programming is almost limitless—fire and burglar alarm systems for the home, shopping and banking services via cable, video correspondence courses, health services, job training—in fact, any data or information that can be transmitted electronically. Two-way audio and video communications between the viewer and the studio are also possible.

At present, the programming potential of cable is unfulfilled. Cable operators have been content to pick commercial television signals off the air and to provide the usual fare of "I Love Lucy," "Julia" and "Bonanza" to viewers who were getting snow on their sets before. The local program origination or "cablecasting" capabilities of the system are virtually unexplored.

Because of its vast programming capacity and income generating power, this emerging new industry is an excellent vehicle for community development. As Frank Thomas of the Bedford-Stuyvesant Restoration Corporation points out:

> The revenue for this system comes out of our community, so what better circumstance for us to be in? We don't have to crash a national market. We don't have to try and compete with manufacturers who've been in business for fifty years. You would think that all of our economic development monies that are existing in various government programs—whether they're special impact programs, or model cities programs—would look very favorably on investments that tend to promote local ownership in this kind of enterprise.

Two workshops have been held to inform local leaders and organizations about the potential as well as the pitfalls embodied in the expansion of cable television into minority communities. In February 1971 more than fifty black mayors, city councilmen, and state representatives convened in Washington, D.C. for a briefing. Then, in June over 100 representatives of community development corporations and organizations attended a three-day workshop on cable communications. A detailed report of the June meeting is included in this publication.

So a beginning has been made. Much more is left to be done. There are few minorities involved at the present time in the important task of informing and directly assisting local communities in establishing and carrying out these development projects. There is no central organization or funds to support the effort.

Yet, participation must be quickly expanded in local communities. Local engineers, businessmen, professionals, media specialists, elected officials and development groups must be brought together to do the remaining work. Most communities are still uninformed and uninvolved in the important decisions affecting cable television franchising now being made at city hall.

Now is the time to enlarge the conversation, the debate, the formulation of public policies, and the ownership and control of CATV in the cities to include brown, black, and red communities and organizations. Information concerning the technology, franchising, local program origination, strategies for community control, community ownership arrangements, business development procedures and prospects, and other pertinent data have been developed and compiled here to facilitate informed action.

This book does not and cannot tell all that one needs to know about cable television. Changes in ownership, technology, and franchise acquisitions within the industry, the formation of new coalitions and interest groups, and changes in regulations by state and local governments are taking place at a rapid pace. Careful study and use of the information provided will enable minority groups to take part in these activities and to establish local development projects.

The FCC held two weeks of public hearings in March 1971 to receive comments and proposals for future CATV regulation. In his remarks at those hearings, Ossie Davis apologized "for appearing so early to present minority demands and interests" in terms of programming, ownership, and control of this new medium. Now is not too early; in fact, minorities are late entrants in the race. They are at least two years behind other participants.

There is still time, however, to act decisively. An informed community can effectively intervene in the political process at the municipal level. Local leaders and organizations must assume the major burden and responsibility in this endeavor.

The editor hopes that the following material and information are useful in meeting these obligations.

Charles Tate,
Editor

CHAPTER 1

an overview of cable television

by ted ledbetter

what is cable television?

Cable television is a method of distributing television signals through a wire rather than broadcasting those signals through the air. Cable was first introduced in remote, mountainous areas where broadcast TV reception was poor. Twenty years ago, private entrepreneurs put TV antennas on mountain tops where they could pick up the signals from distant TV stations. Coaxial cable was strung on poles from the antenna tower into town and individuals homes were connected to this mainline. The system is similar in basic design to the telephone, gas, water and power systems.

Once the cable plant was installed and individual homes were connected, over-the-air broadcast signals were picked up by the antenna tower and transmitted through the cable to local homes. Those who subscribed to this service paid a one-time installation fee of $15 to $25 and a monthly service charge of $5 to $10—a fee and rate structure similar to the utility companies. Thus *Community Antenna Television* was born.

Most of the early systems offered six to twelve channels and it was soon discovered that the quality of picture transmission was far superior to the same pictures received by off-the-air television sets closer to the point of origination. In other words, cable subscribers in an isolated Pennsylvania town many miles from Philadelphia actually got better reception of Philadelphia originated programs than people living in Philadelphia. And in addition, cable subscribers could also receive programs from New York, Cleveland and other cities.

Cable television is able to provide clearer pictures and more channels simply because the TV signals are carried through a cable rather than broadcast over the air. Broadcast signals in close proximity can interfere with each other. Such interference produces "ghost" pictures on black-and-white television and distorts the colors in color TV.

Additional channels are available because cable can use frequencies that broadcast tele-

FREQUENCY SPECTRUM CHART

Figure 1

Broadcast Use	Frequency (megahertz)	Cable Use
	0	
Various Services	6	Sub-Channel a
" "		" " b
" "		" " c
" "		" " d
Fixed & Mobile Radio	30	Cross-Over Filter
Amateur (Ham) Radio		" " "
TV Channel 2	54	TV Channel 2
" " 3		" " 3
" " 4		" " 4
" " 5		" " 5
" " 6		" " 6
FM Radio	88	FM Radio
Aeronautical Radio	108	" "
" " " "	120	Mid-Channel A
Space Research		" " B
Satellites		" " C
Fixed/Mobile Radio		" " D
" "		" " E
" "		" " F
" "		" " G
" "		" " H
Police/Fire		" " I
TV Channel 7	174	TV Channel 7
" " 8		" " 8
" " 9		" " 9
" " 10		" " 10
" " 11		" " 11
" " 12		" " 12
" " 13		" " 13
Fixed/Mobile Radio	216	Super-Channel J
Radio location		" " K
Space, etc.		" " L
		" " M
	240	

vision can *not* use. *The Frequency Spectrum Chart* (p.8) illustrates the Federal Communications Commission allocation of radio bands and channels.

Since signals on cable do not radiate into the air, cable can use *all* of the frequency spectrum, limited only by the cost of very wide band amplifiers. The additional spectrum space available on cable is indicated by the subchannels, mid-band channels and super channels. Note that the mid-band channels are located between regular channels 6 and 7. This means that a twelve channel cable system (Channels 2-13) can be technically up-graded to a 20 or 21 channel system if subscribers are supplied with a switch which converts mid-band channels to frequencies that can be selected on a regular TV set.

There are problems however. For instance, even though the cable can deliver interference-free signals on all channels, strong local broadcast signals can get *inside* many TV sets and produce interference. Secondly, some mid-band frequencies produce interference when mixed with other frequencies. The result of these problems is to effectively reduce the actual cable capacity to 17 channels in certain. communities even though the system can carry 20 in an isolated environment.

Even without the sub, mid and super bands, most cable systems provide increased capacity in individual communities. No city can have twelve broadcast stations without depriving other cities of stations. Baltimore and Washington, for example, can not have TV stations on the same channels without interfering with each other. They are only 40 miles apart and strong transmitters give good signals at least that far. Therefore, the FCC has assigned certain channels to each city in a Table of Allocations.

TV Channel Allocations (VHF)

Figure 2

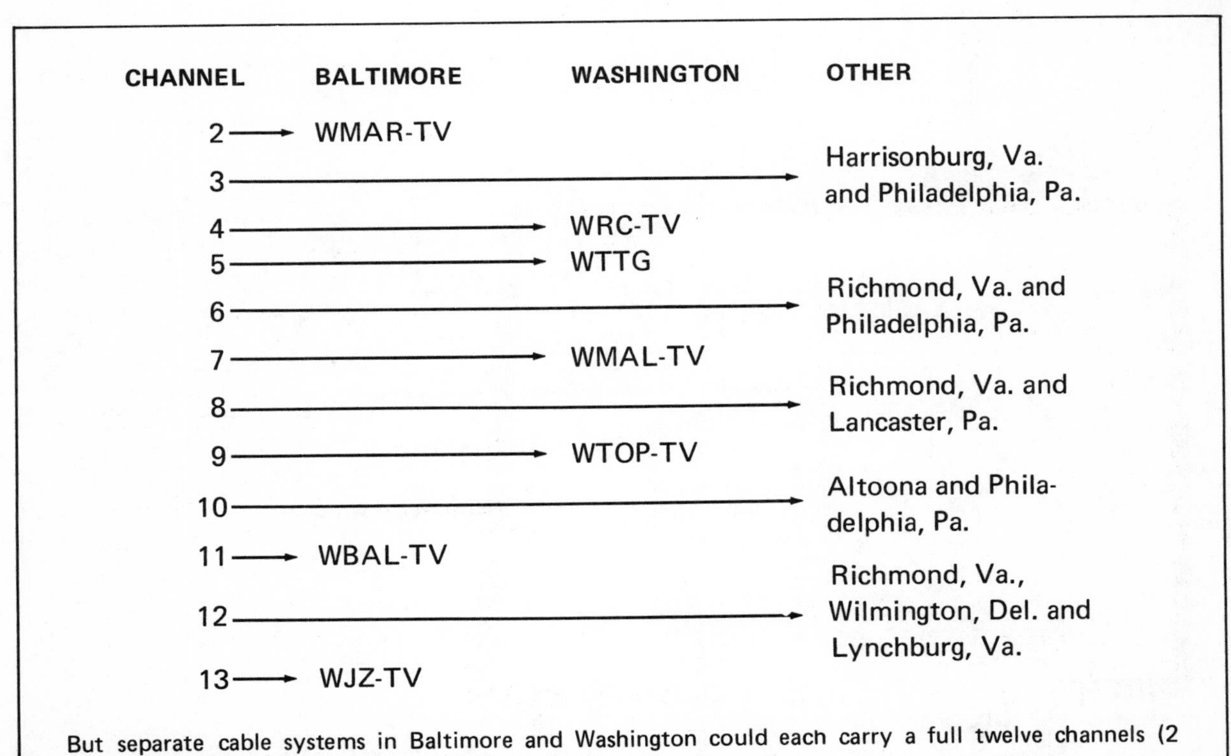

But separate cable systems in Baltimore and Washington could each carry a full twelve channels (2 through 13) without the possibility of major interference.

components of a cable system

A cable television signal receives off-the-air broadcast signals and feeds them through amplifiers and cables to its subscribers. This requires (A) An antenna tower and a head-end, (B) a distribution plant and, (C) house drops and terminals. (See Figure 3.) The head-end (A) consists of receiving antennas, receivers and amplifiers for local broadcast signals. It might also include equipment for translating signals from UHF stations to VHF channels. Head-ends can also include microwave equipment to bring in distant TV signals.

The distribution plant (B) contains amplifiers and trunk cable attached to utility poles or fed through underground conduits like telephone and electric wires.

House drops (C) are taps on the distribution plant for each building. Terminals include connectors, transformers and converters (if necessary) on the subscriber set.

Typical costs are $100,000 for a 12 channel head-end, including microwaves, $4,000 per mile for the aerial distribution plant, $15,000 to $50,000 per mile for underground plant and $40 to $60 per house drop.

For example, Figure 3 also shows a local origination center (D) which is used to produce community cablecasts. Costs for a local origination center can vary from $50,000 to $250,000 for equipment only, depending on the amount of equipment and whether it is black and white or color.

Distribution System
Figure 3

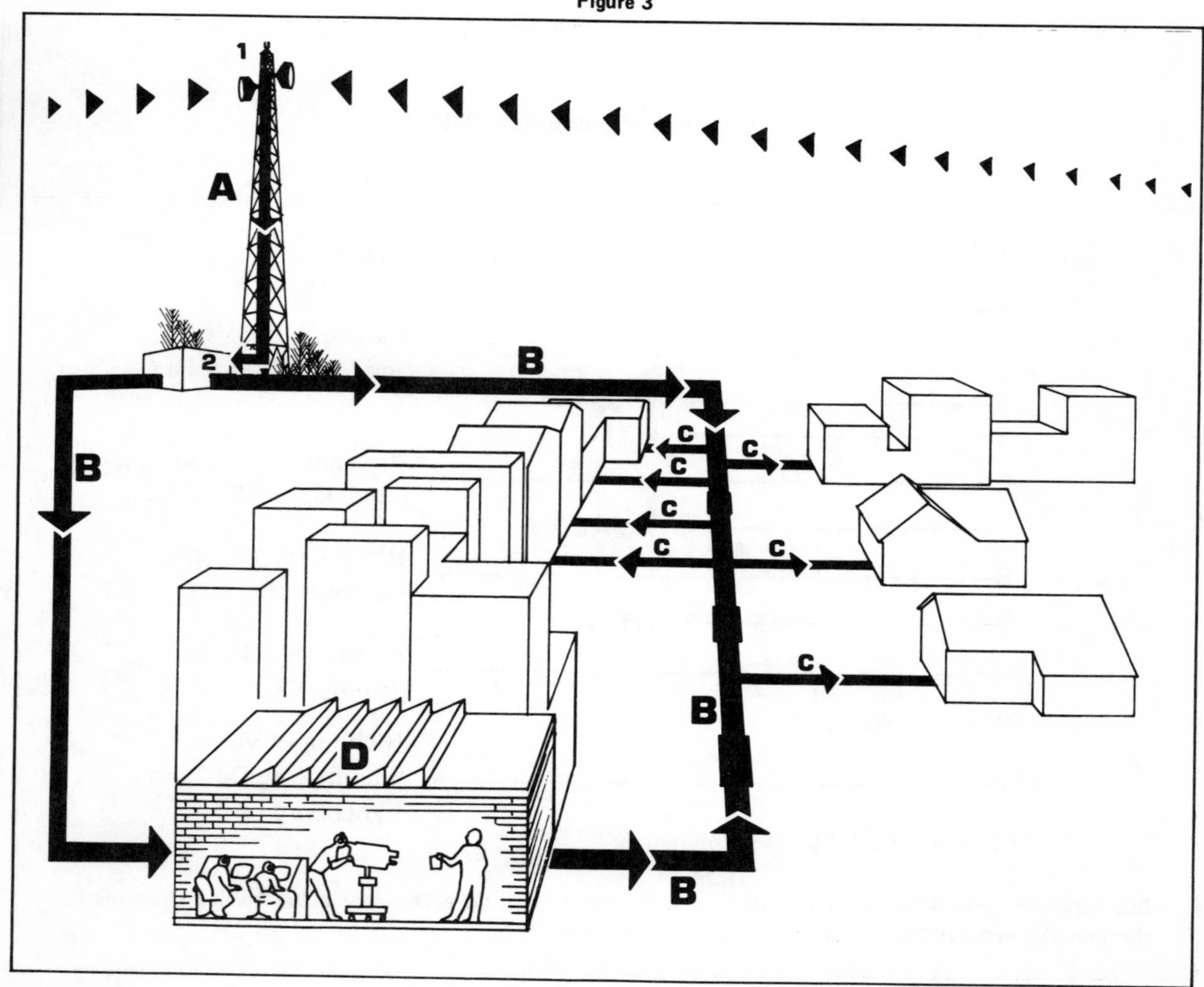

In addition to the basic cable system described above, extra services can be provided. Another possible service is two-way or subscriber response. By using the sub-channels or a second cable, signals can be sent out to subscribers at the same time that signals are returned to the head-end or origination center.

A third additional option is the multi-cable distribution plant. Although it costs more than a single cable system, a multi-cable system can multiply the number of available channels as many times as there are cables. In Reston, Virginia, for example, subscribers have a small switch box on the back of their TV sets for selecting between cable A and cable B. This two-cable distribution plant provides 24 channel capacity in Reston, 12 channels on each cable.

There are other, more futuristic options which are described later, but there are two fundamental notes to make about any cable system.

1. **Channel capacity** is technologically unlimited. Across the country, 20, 40 and 60 channels systems are now under construction.

2. **Local distribution** is inherent in cable. The signals go through the cable and can be received only where the cable goes. Therefore, the order in which franchises are awarded in each city, the order in which the cable franchisee elects to install cables within each part of the city, and the rate of cable plant construction determine who the consumers will be.

the CATV industry past and present

The cable television industry started in 1948 in the hills of Pennsylvania and Oregon. From 70 systems with 14,000 subscribers in 1952, it has grown to more than 2,500 systems in 4,400 communities with about 14 million viewers. Total industry revenue is approximately $300 million. But cable serves only 7% of the total population and only 1.6% of the TV homes in major metropolitan areas. And, only 350 of the existing systems have more than 3,500 subscribers.

Within the next ten years, cable penetration is expected to reach about 70% of the total TV homes and to produce industry revenues of $4.4 billion. One reason for these optimistic projections is that, in addition to the 2,500 operating systems, another 2,300 cable franchises have been granted by local communities and an additional 2,600 applications are pending. Most of the pending franchises are in urban markets where the FCC is expected to remove restrictions on cable development by early 1972.

Throughout its history, cable has been a technological and political unknown. At first it appeared to be all good because it provided TV service to homes not reached by broadcast stations; the subscriber, broadcasters and cable operators were happy. But as cable systems moved into metropolitan areas, broadcasters began to fear that, by importing distant signals, cable would "fractionalize" their audiences.

Other interested groups, including telephone companies and newspaper owners, began to see new, unwanted competition. And they acted in two ways to meet this competition. They pressured the FCC to stop ignoring cable and put restrictions on the importation of distant signals in the Top 100 markets. They also started building and buying their own cable systems. Today, approximately 50% of the cable industry is owned by other media interests, including broadcasters (36%), newspapers (8%), telephone companies (6%). Among the largest multiple-system operators (MSO) are Cox Broadcasting, CBS, Time-Life, Foote, Cone & Belding, United Utilities, General Tire and Rubber, General Electric and Westinghouse. Other MSO's include CATV equipment manufacturers and small savings and loan and finance companies.

Perhaps the most interesting cable company is the largest—Teleprompter (TPT). TPT serves more than 500,000 subscribers on 153 individual systems around the country. Teleprompter's cable franchises include Manhattan, Newark, Gary and Trenton. Through Theta Cable, which TPT owns with Hughes Aircraft Corp., it is the largest operator in the Los Angeles area. TPT's best known shareholders, Howard Hughes and Jack Kent Cooke, own a total of 668,000 of the 3.2 million shares outstanding.

The FCC is moving to prevent cable from affecting the economic health of broadcasting, and, to a small degree, has placed some limits on cable ownership. TV networks must divest themselves of cable ownership. TV stations and telephone companies can not own cable systems in their own markets and newspaper cross-ownership is being questioned.

The present structure of the cable industry is quite different from the small systems of fifteen years ago. With the anticipated move of the cable into the big cities, it will change even more. Unless creative financial and political strategies are developed, only the major corporations will be able to invest an average of $20 million for construction of a cable system in each major city.

GROWTH OF THE CATV INDUSTRY
(as of January 1 of each year)

Figure 4

YEAR	OPERATING SYSTEMS	TOTAL SUBSCRIBERS
1952	70	14,000
1953	150	30,000
1954	300	65,000
1955	400	150,000
1956	450	300,000
1957	500	350,000
1958	525	450,000
1959	560	550,000
1960	640	650,000
1961	700	725,000
1962	800	850,000
1963	1,000	950,000
1964	1,200	1,085,000
1965	1,325	1,275,000
1966	1,570	1,575,000
1967	1,770	2,100,000
1968	2,000	2,800,000
1969	2,260	3,600,000
1970	2,500	4,500,000

Source: *TV Digest*, 1970-1971.

the potential of cable

The potential of cable is basically two-fold. First, it is a means of *communication*, supplying the kind of information and more diverse programming at a lower cost which is responsive to the needs, desires and aspirations of all people.

Secondly, cable is a *business* with great potential for cash flow and jobs.

Cable TV As Communications

Cable could affect almost every type of social pattern in existence. Areas such as:

1. Work Activity—speeding information flow (e.g. mail) by facsimile
2. Data collection—increasing computer accessibility and interaction
3. Politics—providing low-cost access, even for local candidates
4. Transportation—Video monitoring and computerized control of traffic
5. Public Safety—providing emergency communication for combatting crime
6. Education—extending access and availability (including 2-way response)
7. Entertainment—increasing variety and decreasing dependence on mass audiences
8. Consumer—providing consumer education
9. Marketing—offering demonstrations and remote retail shopping
10. Community Development
    a. Meetings of civic associations and model cities groups can be aired.
    b. Telecasts of hearings that pertain to the community's particular interests can be broadcast.
    c. Cohesion can be created through the showing of entertainment and discussion programs that originate from the neighborhood.
    d. There can be an information center for efficiently giving news and instructions to community residents.

Commercial television is the most pervasive educational vehicle in our society today, despite the fact that it is seldom perceived as such. Imagine CATV as a vehicle for producing an educated population—and the subsequent effect of this newly sophisticated electorate on our political system. Cable offers the possibility of widespread personal video communication among people with similar goals and objectives.

The potential for greater freedom of spirit and mind is staggering, as is the possibility for social recidivism and intellectual stagnation. It is awesome to speculate about some of the related developments of this technological achievement.

Because cable will offer more variety in entertainment, educational and business type programming, more people will stay home to watch their TV sets and will be less inclined to attend movies, live performing arts productions, sports attractions, community and political rallies, educational classes, business seminars and even church. Proposed two-way communication systems will reduce the need to leave home for shopping trips, minor medical attention, or repair services. In ten years, children may not attend school (a physical building) except occasionally. They will select educational programming in their home and choose a favorite teacher. This method may provide a better education; but who will determine what is a better education?

The Business of Cable TV

As an industry, cable is expected to generate $4.4 billion in income by 1980. It could create over a million *new* jobs. For example, there are less than 900 authorized TV stations in the U.S. If each of the 5,000 projected cable systems has an origination studio, they will require six times as many cameramen, technicians, clerical workers, etc., as are employed by the broadcast TV industry today. But there is more: cable systems need installers, linemen, repairmen, managers and others and they will require goods (amplifiers, cable, office supplies) and services (accounting, printing) from other companies.

A typical large cable system (500,000 population community) could cost $15 million to build, but it would gross $10 million in revenue after five years. Since the system has a physical plant, it can use depreciation directly or sell it to limited partners. There is also the potential for cable spin-offs, such as pay-cable program syndication, computer and security services, all of which could be operated by independent businessmen with channels leased from a cable customer.

In summary, cable has the potential to help or hurt communities. It can help community organizations get access to the media and by providing better educational facilities. It can hurt by reducing personal contacts to a minimum, by eliminating half of the postal jobs in this country, and by restricting public access, program diversification and community responsiveness. It offers new job and business opportunities, but these opportunities could turn out to be "white only." It could provide increased public safety: but who wants the television equivalent of "no-knock" police power?

CATV has the potential to help minority people solve many problems and achieve a new plateau in self-determination and self-fulfillment; it could also produce the Orwellian nightmare long before 1984.

CHAPTER 2

community control of cable television systems

by charles tate

introduction

The prospect of cable television expanding into the metropolitan centers throughout the country has caused a major power struggle among wealthy and influential investors, local governments, and various well-financed, vested interest groups. The central cities within the Top 100 markets have already become hotly contested franchise areas. If blacks, Chicanos, Puerto Ricans and other racial minorities who reside in the central city ghettoes lay claim to those systems that will operate in their communities, ownership and control by outsiders could be seriously challenged and prevented.

Since blacks are a dominant minority in most of the major cities, their action or lack of action will greatly influence the manner in which government and the private sector respond to minority efforts and demands for participation in the ownership, control and programming of these systems. Although this chapter focuses on the problems, potentials and the prospects for participation by blacks, much of the following discussion is broadly applicable to Puerto Rican, Indian, and Chicano communities.

communications technology and community control: confrontation and challenge

The long and difficult struggle by blacks to extricate and liberate themselves from the exploitative, oppressive, and racist conditions in America has involved many tactics and programs. Black self-development and self-determination efforts have consistently emphasized, however, the necessity for control of those public and private institutions that operate within their communities. Conflicting ideologies have not diminished the interest, enthusiasm, or efforts to achieve these goals. In fact, these differences have strengthened the determination to overcome the obstacles that thwart success.

Pan-Africanists, integrationists, separatists, and black nationalists advocate community control of community institutions. DuBois, Booker T. Washington, Marcus Garvey, Malcolm X, Stokely Carmichael, Elijah Muhammad, and Huey Newton are in agreement on this issue.

The increased urbanization and concentration of the black population in the central cities has given additional impetus to this historic movement for community control. Contemporary efforts to gain control of economic, political, cultural, and social institutions are a product of the civil rights movement. Once the apartheid system of racism of the white South was openly challenged, blacks trapped in the urban ghettoes in the East, North and West were motivated to challenge the more subtle, but equally entrenched and oppressive systems of racism that operate in urban centers. The battle for social justice and self-determination quickly expanded from Little Rock, Birmingham, and Selma to Chicago, Brooklyn, Oakland and Watts.

Because of the more sophisticated and complex structure of racism and decision-making in urban governments, community control has become the dominant theme in the struggle of urban minorities for social justice. Community control challenges white control of those institutions that operate in and serve predominantly black communities. Through these institutions, whites exercise control over the resources needed for local development.

The public school system, poverty programs, unions, police departments, welfare agencies, United Appeal and every form of urban-based institution and organization controlled by whites are now being challenged. The urban oppressed are demanding jobs and economic benefits as well as a controlling voice in the policy-making functions of those institutions, agencies and programs operating in their communities. The results thus far are small, but important changes are taking place in the degree and quality of minority participation and control of local institutions, organizations and programs.

White resistance, rooted in fear and racism, has obstructed and hampered the community control movement. Economic sanctions, assassinations and political repression have been used to control, isolate and divide black communities.

Community leaders and organizations are now faced with a new challenge in their efforts to achieve community control. *Cable television, a futuristic communications system ideally suited for community control and local programming, is on the verge of broadscale expansion into the cities and ghetto communities.* This development could provide the leverage needed by local communities to achieve a much greater degree of independence and self-determination or it could seriously weaken the movement. Cable television will have a decided impact one way or the other. Its importance and its potential as a social, cultural, economic and political force cannot be ignored.

In this age of technology, cable television is a super technology—a synthesis of radio wave electronics and computer technology. It's a powerful illustration of Alvin Toffler's theory that technology feeds on itself and that the elapsed time between technological innovations has been so drastically reduced that man is already living in the future.

Cable television may be the last communications frontier for the oppressed. Yet, most community leaders and organizations know nothing about this revolutionary communications technology and the plans underway to install sophisticated video systems in homes, schools, hospitals, health centers, courtrooms, police stations, banks, fire stations, supermarkets and department stores.

Significant social benefits or accelerated community disruptions could result. Reinforcement and strengthening of cultural integrity or cultural genocide are equally possible. A reduction in political repression or a substantial increase in fascist style police surveillance, harassment, and oppression could be realized.

All of these results are possible because of the vast signal carrying capacity of cable systems. Existing technology can provide up to 60 interference-free channels for the transmission of both audio and video programs, as well as the two-way transmission of video, audio and computer type data.

The community control and development potential is in this copious programming capacity and the economic structure of cable systems. Subscribers rather than advertisers are the bread-and-butter of these enterprises. A monthly fee is charged to cover installation and maintenance costs for connecting individual homes and apartments to the cable—an arrangement very similar to the telephone system.

Confronted with minimal regulation by the federal government and local municipalities, a small group of entrepreneurs has realized vast profits from cable in small towns and rural areas. Major investors and media owners are now pushing to bring cable to the major cities and the central city ghettoes.

It is estimated that cable television will be a $4.4 billion industry creating over a million *new* jobs by 1980. Other projections indicate that a completely wired, interconnected national broad band communications system would create an $80 billion service industry.

Naturally, cable operators are anxious to expand operations into the cities, especially the big cities. They are ready to reap the profits from residential subscribers and the anticipated market that will develop for a wide variety of business spin-off services. The Federal Communications Commission has blocked them with various rules at the insistence of the influential broadcast lobby. Power is beginning to shift in the opposite direction. The FCC recently proposed a new set of rules to Congress and the President that would remove the basic restrictions early in 1972.

A major power struggle is underway among broadcasters, cable operators, the FCC, Congress, the Administration, newspapers, publishers, motion picture producers and allied media interests, and a variety of professional groups and associations. All are jockeying to influence the development, expansion, use and control of cable systems in the cities. The stakes are high.

Because there is great power and profit potential in the ownership and control of this medium, the oft-repeated "rip-off" by big business interests for private gain at the expense of the public interest is taking place once more. If it succeeds, it will stifle the diversified, highly specialized, local programming potential of CATV and prevent local control and community development. Diversified public and private ownership offers the best assurance that social benefits rather than social disaster will be the end product of this new media. Concerted action by minority groups at the local and national level can bring positive results.

Among those public groups engaged in the policy debates, most advocate a regulatory arrangement that would guarantee minority groups and individuals public access to one or more channels on a free or minimum charge basis. A regulatory scheme requiring uniform toll rates similar to the rate system used for long distance telephone operations has been suggested.

Access is extremely important to minority communities, but it does not go far enough. Access alone will not provide the measure of control required over capital, labor and technology, to stimulate and sustain economic and social development of ghetto areas. Ownership and control must be achieved to meet this objective. A sizeable portion of the income and profits from CATV in the major cities will come from minority subscribers, particularly blacks. Unless these systems are controlled by the communities served, the resources urgently needed for development will be lost.

If this proposed "access" policy were applied across the board, there would only be white-owned businesses in every sector; a conclusion not only at odds with the goals of self-determination, but one certain to render blacks and other minorities even *more* powerless and dependent. If it is adopted as the public policy for minorities in the field of cable communication, it is certain to increase the power of the white business community, utilizing minority revenues as a subsidy. In other words, ghetto communities will be placed in the position of "paying" for their powerlessness and economic dependency.

Access and community ownership and control are not mutually exclusive or antagonistic. Regulated public access and community ownership and control are equally desirable objectives. In fact, ownership and control may provide the only safe guarantees that access will be accorded to minorities on a nondiscriminatory basis.

Why should the community settle for the privilege of buying time or being allowed free time on a system that it supports through its monthly fees, when it has the right and opportunity to operate and control the system and serve as the administrator of the public access guarantee?

The continuing oppression and exploitation of blacks and other racial minorities is directly related to their lack of control over indigenous institutions and resources. For example, it is estimated that the annual disposable income of blacks alone is about $40 billion. On the other hand, a current survey of minority businesses by the Bureau of the Census revealed that receipts of *all* minority businesses in 1969 were less than *one percent* of total U.S. business receipts—a meager $10.6 billion. Minorities own less than five percent of the total businesses in the country, and most are small retail and service operations with fewer than five employees.

Minority ownership and control of cable television systems could dramatically alter this situation over the next five to eight years. There are approximately twenty-five cities with black populations in excess of 100,000; eight of these have populations in excess of 200,000; five have populations over 500,000; and two have populations over 1 million. After five years a cable system with 10,000 subscribers would generate revenue of approximately $500,000 annually and a system with 20,000 subscribers would generate up to $1,000,000 in revenue annually. Obviously, many of these communities could support several cable systems. Six cable districts have been proposed in Washington, D.C. where the black population is over 500,000.

Are there real opportunities for community ownership and control? Can minority-owned systems serving local communities be financed? Can they become successful economic enterprises?

Will ghetto residents pay $5 a month for more channels and better quality pictures? What are the dangers to minority communities associated with the growth of cable television systems?

Answers to these questions require a more careful and detailed examination of the politics and the economics of CATV development, ownership patterns in the industry, the market potential of urban black communities and the kinds of corporate arrangements that are compatible with the goals of community ownership and control.

the new frontier

From a few isolated cable systems in small communities and rural areas of the U.S., this new industry has aroused nearly every power bloc and organized interest group in the country. Their excitment centers around three aspects of CATV: (1) profits, (2) the vast signal transmission capacity of cable, and (3) the imminent expansion of cable television into the cities and major metropolitan centers.

Cable has turned out to be an extremely profitable venture for investors. Multi-millionaires have been created by the cable television industry. A classic example is Milton Shapp, incumbent governor of Pennsylvania. Mr. Shapp invested $500 in a cable venture in the early fifties. When he first ran for governor of Pennsylvania in 1966, Mr. Shapp sold his CATV interests for approximately $10 million.

Cable systems seem to offer unlimited opportunities for making money. First of all, cable operators have avoided programming costs by retransmitting programs produced by broadcast television. This air piracy was upheld by the U.S. Supreme Court. Accelerated depreciation schedules enable operators to gain the benefits of a tax loss while increasing the book value of their investment. The practice has been to depreciate systems over a five to eight year period even though equipment life is actually twenty years. The results—a guaranteed paper loss during the first three to five years of operation. This loss is allocated on a pro rata basis to investors who then claim the loss on their individual tax returns. Meanwhile, the cash deductions made for depreciation before taxes are available for use to expand the system or purchase new ones. Hence, the assets of the system and the book value of the stock are increased. When the system is totally depreciated, it may be sold and the entire process can be started all over again.

New installations and extra fees for extra services also increase system revenues. Like the telephone system, each time a cable outlet in the home is moved a fee is charged. Stock ticker tapes, weather information and similar forms of static programming may be provided to individual subscribers for an extra fee. And generally, subscriber fees have been raised with little opposition from the local commission or council, whose members may be investors or recipients of gratuities from the cable companies.

The vast signal transmission capacity of cable television is further cause for excitement among power blocs and organized interest groups. The "economics of scarcity" common to over-the-air broadcasting can be eliminated by the enormous channel capacity of cable. Early systems provided up to twelve interference-free channels and those now under development will offer from 24 to 60 channels. This abundance means the general public can now afford video programming for a wide range of purposes; e.g., education, community meetings, and information programming concerning health, jobs, and legal matters to mention a few. Private access, similar to the telephone system, is possible using rate schedules like those for long distance calls.

Cable technology has a potential, however, that goes far beyond increased channel capacity. Two-way communications, home computer terminals, home banking and shopping services, transmission of mail, fire and burglar alarm systems, piped-in music for each home, and other 1984 style communications services can be provided over the same cable that transmits the video signal. "Cable television" is a misleading term, "cable communications" more accurately defines the technological parameters of this new medium.

With the highly probable interconnection of systems between cities by domestic satellite within this decade, the communications prospects of cable technology are genuinely "mind blowing."

The *third* and perhaps most crucial factor generating the growing interest and controversy over this medium is the introduction of cable systems into the major central cities and metropolitan areas. Densely populated cities and metropolitan centers comprise the Top 100 markets for commercial broadcasting and its advertisers. These centers contain the concentrated buying power for commercial products that manufacturers and broadcast networks compete for. It is the broadcast networks' demonstrated ability to get 40 to 50,000,000 viewers for programs offered in the prime viewing hours of 8 to 11 p.m. that enables them to charge extremely high rates for commercial advertising.

The lucrative rate of return, however, has not caused the industry to adopt a particularly high level of conscience. Mass audience programming, exorbitant costs for television air time, the hard selling of commercial products (including the irresponsible practice of promoting unsafe products) are all an integral part of broadcast television. The scarcity of over-the-air transmission space has created a small but extremely powerful group of owners who have demonstrated a greater interest in profits than public service.

Because of the power and influence of the broadcasting lobby, the FCC adopted a series of rules beginning in the early sixties that prevented cable operators from importing distant signals into the Top 100 markets. The "freeze" effectively stopped urban cable expansion. Distant signals—programs that the local viewing audience cannot normally pull in with roof-top antennas—make cable systems highly profitable. Deprived of this advantage, in the Top 100 markets, owners and investors simply went where the money was—small towns and the suburbs.

In recent pronouncements, the FCC has made it clear that it plans to lift the "freeze" on distant signal importation early in 1972 in the Top 100 markets. The prospective thaw has created a phenomenon not unlike the California gold rush. Wealthy investors are buying cable systems and franchises. City governments, hungry for revenue, are issuing long-term franchises in return for a percentage of the profits. The city "take" from gross revenues ranges from 2% to 30% across the country.

White, middle-class professional groups and associations are lobbying for public policies, at local, state and national levels which will give them a strong voice in the operation of local systems in the areas of public access channels and local programming.

The Sloan Foundation has financed a $500,000 national commission on cable television to study and recommend public policies regarding future uses.

The Ford Foundation has financed a series of studies by the Rand Corporation on cable television including a study by Rand in Dayton, Ohio. The Mitre Corporation is conducting a study in Washington, D.C. with a grant from the Markle Foundation.

The Ford and the Markle Foundations are planning a $1 million per year National Cable Information Service to provide information and technical assistance to state and local governments on franchising and overall use of cable communications within local communities.

If the present trends continue, minority communities will be excluded and disenfranchised. White capitalists who own and control the major print and electronic media systems in America will own and control the cable communications industry, including the systems that serve black communities. Fifty percent of the cable industry is already controlled by other media owners. Broadcasters own 36%, newspapers 8%, and telephone companies, advertising agencies and motion picture companies 6%. Further, there is a rapidly developing concentration of ownership within these groups. Ten companies now control 52% of the industry. The top ten, in order of ranking are: Teleprompter, Cox Cable, American Television and Communications, Tele-Communications, Cypress Communications, Viacom, Cablecom-General, Time-Life Broadcasting, Television Communications and National Transvideo.

The white middle class that manages and operates major educational, social and cultural institutions (i.e., schools, colleges, universities, foundations, theaters, museums and churches) is actively vying to dictate public programming policies for cable systems, including those serving black communities.

These two groups—white capitalists and the white middle-class intellectuals, managers and technocrats who have worked so effectively together in controlling and operating everything from the Pentagon to the poverty programs (at a handsome return to each group) —are now moving toward an accomodation of interests in this new communications field. If this act is consummated, the promise and the potential of CATV as an instrument for empowerment and development of underdeveloped ghetto communities will be seriously diminished, subverted, and perhaps entirely lost.

---

what you see is what to get

In view of these characteristics and dynamics, what are the opportunities and obstacles to community control and development associated with the expansion of CATV into the cities?

The requisite conditions for community control of resources and development are mass mobilization and unified action.

For example, urban renewal programs provided a significant opportunity for unified action by varied interest groups within urban black communities. Many ghetto communities united to stop these projects because of the insensitive treatment of residents by urban renewal agencies and the disruption of the community for the benefit of white profiteers. Serious attacks were made against the traditional system of planning and decision-making from the local to the national level. Blacks demanded and secured important concessions affecting policy-making, jobs, and other aspects of the urban renewal process.

Cable television is a better vehicle for achieving sizeable gains in community organization, unification, control and development. Several factors support this conclusion. First, cable television systems are not presently installed in black communities and central cities. Therefore, no entrenched interest group or power bloc can claim public protection for its investments. Second, franchises are issued by local, municipal governments, and the FCC has recommended the continuation of this process. Third, installation requires the actual stringing of cable on poles or the laying of cable underground along the streets of the ghetto. Individual hook-ups must then be made from these trunk lines to homes and apartments, and outlets must be installed within these living units. Fourth, black communities are a substantial segment of the urban subscriber market. Fifth, the great potential of cable in technology, economics and the power of mass influence is ultimately tied to cablecasting or local programming origination. Sixth, cable will be used in a wide variety of applications, apart from entertainment programming. Education, health, welfare, safety, crime prevention and police operations are a few of the likely uses.

Viewed together, these factors reveal significant opportunities for community participation and the imperative need for community control. Each of the listed factors will require a series of crucial political decisions at the local level. How will CATV be regulated? How many franchises will be awarded? Will a single franchise cover both system management and system programming? What are the qualifications for franchise applicants? How will franchise fees collected from CATV enterprises by the local government be used? Who will own and operate the systems? Who will determine the program content? Who will install the systems? Who will decide on the areas to be served? These are political issues that will be decided with or without community participation, but the options for black communities are still open. How long they remain open will depend on the initiative, ingenuity and determination of community leaders and organizations.

Whether or not a franchise has been awarded, broadscale community participation is possible. Early involvement in the franchise process is crucial. Local franchising involves several steps. The commission or city council usually adopts an ordinance giving the political body the legal powers to regulate CATV within the community–including procedures for awarding franchises. The ordinance may stipulate that public hearings must be held prior to franchise awards and that public notice must be given regarding the period established for filing franchise applications by interested parties. The ordinance may further state that multiple franchise awards will be made for various geographical areas within the city or county.

Community organizations should view the entire franchising process as an area of vital interest to their constituencies. Ideally, disenfranchised black communities should be consulted by the local government and included in the discussion and development of the ordinance and all other regulatory aspects of CATV systems for their communities. Needless to say, that is not happening. Local politicians, who are not well informed about CATV, have been selling the "communications birthright" of minority communities to the highest bidder.

Community participation should begin with a systematic, factual determination of the status of the franchising process within the local government. The city attorney or council should be contacted for this information. If no ordinance has been adopted or franchise awarded, action should be taken to establish procedures for community inclusion in the policy–making process.

If an ordinance has been adopted and/or a franchise awarded, a detailed review and evaluation should be made to determine the provisions made for community participation in the monitoring, control, programming, and ownership of the system that serves them.

Most CATV franchises are nonexclusive agreements between the city and the cable operator. Thus, community groups may organize their own company and apply for a franchise covering the same territory as previously awarded.

As a last resort, it may be necessary for the community to exercise its veto power over those CATV projects that disenfranchise blacks. Legal and other forms of protest actions may be required to achieve community participation in the policy-making discussions and the achievement of community control and development objectives.

---

opportunities for community development

Cable television provides a substantial opportunity for urban minority communities to develop and control the most powerful cultural and social instrument in their communities. It can also provide a viable economic base and political leverage for power-deficient communities.

A partial listing of the wide variety of program uses will give some idea of the development possibilities:

Educational Uses

- video correspondence courses
- special education programs for unskilled workers, housewives, senior citizens and handicapped persons
- home instruction for students who are temporarily confined
- adult education programs
- exchange of video-taped educational programs with other schools; e.g., science, travel and cultural programs
- interconnection of school systems to facilitate administration, teacher conferences, and seminars
- greater use of computerized testing and grading—thus giving teachers more time for individual instruction

Health Uses

- interconnection of medical facilities (private offices, clinics, hospitals) to provide a wider range of consultative services to patients on an emergency or nonemergency basis—especially those without means of transportation
- wide dissemination of preventive medical and dental information to the community
- information programs concerning sanitation, sewage, rat control, garbage control and similar problems

Legal and Consumer Uses

- listing of substandard and abandoned housing
- review of leases, agreements and installment contracts
- discussion of labeling, marking, pricing of food, drug, clothing, automobile and other consumer products
- establishment of a "hotline" in legal aid and consumer protection agencies to provide immediate notice of fraudulent and exploitative practices
- use of video-taped records and depositions in non-jury cases

Safety Uses

- installation of fire emergency and burglar alarm systems in every home. (These systems can operate over the same cable that brings in video signals.)
- automatic gas, water, and electric meter readings
- rumor control
- disaster and emergency warning systems

Cultural and Entertainment Uses

- Minority-owned cable systems in the Top 50 television markets alone would provide a major market as well as a distribution system for professionally-produced films, plays, concerts, sports events, talk shows and every other form of artistic, creative and intellectual expression. There is no shortage of professional talent in the community—only the lack of a mass-based communications and distribution system controlled by blacks. A community-owned and controlled distribution system could have promoted the Ali/Frazier fight. The white promoter of the fight, Jack Kent Cooke, is a major CATV owner.
- production of a black history series from the voluminous materials written by DuBois, Hughes, Malcolm X, Cullen, Woodson, Bennett and thousands of minority historians, politicians, writers, poets, and leaders who have prepared records of their people's struggle. Such a series could now include an oral history of important historical events by elders of the community.

is there gold in the ghetto?

The economic potential of CATV for minority communities should not be minimized or overlooked.

The urban ghettoes in America comprise a compact, differentiated and lucrative market for cable television—a conclusion that is supported by the phenomenal economic success of soul radio stations. This fact has not escaped the attention of Teleprompter, Time-Life and other white entrepeneurs who are scrambling for ownership and control of cable systems in every large city.

Cable is uniquely suited to serve as a vehicle for economic development, because it is a subscriber-supported system. If an adequate number of households in the community purchases the service, sufficient income can be derived to maintain the system and to produce a profit.

Most community-based enterprises that depend on black customers (with the exception of white-controlled, high risk, illegal operations like the numbers and narcotics rackets) are small, marginal operations. Further, the minority businesses that do exist must compete with white-owned enterprises that enjoy the benefits of volume buying, better locations, off-street parking, lower insurance, credit services and other factors that make the difference between economic success and failure.

It is worth noting that cable systems do not require these advantages to break even or make a profit. Think of the telephone system again. Its services are provided by wire. You don't need a good location, off-street parking or any of the other externalities generally required for retail businesses to maintain a viable telephone system. You take your service to the customer's home and he pays a monthly fee. The same applies to cable television.

Cable television is inherently a monopolistic enterprise. Although it is possible, it is highly unlikely that there will ever be more than one cable system serving a given community. Therefore, a black-owned system serving the entire inner city or just the black community would have a captive market—just as a white-owned system serving a ghetto community would have a captive market. (Soul television is not a remote possibility.) The point is that a community-owned system would not have to compete with white-owned systems downtown or in the suburbs as minority-owned grocery stores, restaurants, hotels, motels, clothing stores, drug stores, et al., must.

The economic success of a cable system is determined by the following factors:

1) the total population base
2) housing density
3) market saturation
4) the number of channels available
5) the quality of reception for existing over-the-air broadcast television signals

Can a community-owned system serving the inner city, or only the black community, thrive? The prospects look good.

One indicator is the list of Top 50 black markets for television recently published by *Black Communicator*. *(see page 26)*

These statistics show there is a substantial black population base in the major cities. The number of households* in each city is even more significant. Considering that 95 to 98% of these households have television sets, the subscriber market for CATV programming is most attractive.

These factors must be considered in terms of the hard economics of capitalization, construction and operating costs, however, to present a realistic picture.

A CATV plant consists of an antenna tower with preamplifiers; a head-end—which is an electronic control center comprised of a wide variety of amplifiers, signal processors, filters and traps; a trunk line coaxial cable fitted

---

*The household data is for the total metropolitan area. To arrive at an approximate number of black households in each city, the total black population should be divided by four.

with amplifiers; feeder lines that are also fitted with amplifiers; tap-offs; housedrops; outlets inside the home; and a small converter/transformer that matches the coaxial cable to a conventional TV set. The bulk of the plant is the coaxial cable that provides the signal distribution network. The cable plant thus accounts for the major cost of the system. The size of the plant depends on the size of the community. As a result, plant costs as well as operational costs are also proportional to the size of the system.

Plant operating expenses consist of salaries, pole rental, insurance, office rent, promotional expenses, taxes, heat, light, power and other items common to most business operations.

Capital outlay necessary for a CATV system can be roughly figured in the following way:

a) distribution system—multiply the miles of system times the cost per mile. A working figure is $4,000 per mile for an aerial system.[1] Estimates for underground installation, vary from $15,000 to $50,000 per mile but many cities have telephone and power lines installed on poles rather than in underground conduits.
b) head-end and tower—towers for most operating systems average about 500 feet. Tower costs run about $25-$35 per foot. Additionally, a tower has a separate antenna, amplifier and various filters and traps for each over-the-air signal. A rough estimate of the cost per signal is $22,000 to $30,000. Tower site preparation, head-end equipment and head-end building (8 x 8 x 12 insulated and air conditioned) comprise the bulk of the additional costs.
c) extra charges—use local estimates for office remodeling, office equipment, vehicles (cars and trucks) and telephone and power pole make ready. Make ready estimates can be secured from local power and telephone companies.
d) housedrops—roughly $10 per housedrop and $5 per housetap.

In a 1968 report, Drexel Harriman Ripley, an investment research firm, developed an economic profile of a twelve channel system serving 10,000 homes in a community of 30,000 to 40,000 people.[2] Housing density was estimated at 100 homes per mile.

100 miles of cable at $4,000 per mile	$400,000
head-end equipment	75,000
legal fees, promotional and other costs	85,000
	$560,000

Estimating market saturation (subscribers) at thirty-five percent by the end of the second year, the report projected an annual accounting for the system as follows:

Revenues (3500 subscribers at $63 each)[3]	$220,000
Operating Costs	124,000
Operating Profit	96,000
Depreciation and Amortization (10-year write-off straightline)[4]	56,000
Interest (on $400,000 at 7 ½%)	30,000
Pre-tax income	10,000
Taxes obviated by tax loss carry forward	. . . .
Net income	10,000
Cash flow (normal accounting basis)	66,000

(Note: Hookup charges range from $10-$20 and multioutlet installation charges are usually $1.00 to $1.50 per outlet.)

Assuming that $475,000 of the initial capital outlay was borrowed, the return on equity of $85,000 is 12% at the end of two years and the system is generating a cash flow of

[1] This figure could vary from $4,000 per mile to $25,000 per mile, or even $50,000 per mile for underground construction.

[2] These figures should be used for comparison only as they are not based on urban density.

[3] Includes hook-up fee and multi-outlet charges.

[4] Depreciation is accelerated for tax purposes.

TOP 50 BLACK MARKETS

CITY	BLACK POPULATION (city only) 1970 U.S. Census			TV MARKET RANK TOTAL METRO AREA (FCC)		CABLE TV PENETRATION TOTAL METRO AREA (NCTA)	
	Rank	Total Blacks	%	Rank	Households	Subscribers	Systems
New York	1	1,666,636	21.2	1	3,353,000	57,281	9
Chicago	2	1,102,620	32.7	3	1,613,000	337	1
Detroit	3	660,428	43.7	5	866,000	0	0
Philadelphia	4	653,791	33.6	4	1,339,000	18,048	10
Washington, DC	5	537,712	71.1	9	631,000	0	0
Los Angeles	6	503,606	17.9	2	2,015,000	63,171	25
Baltimore	7	420,210	46.4	14	406,000	3,500	2
Houston	8	316,992	38.9	15	404,000	0	0
Cleveland	9	287,841	38.3	8	764,000	15,000	2
New Orleans	10	267,308	45.0	31	269,000	0	0
Atlanta	11	255,051	51.3	18	369,000	3,560	2
St. Louis	12	254,191	40.9	11	543,000	0	0
Memphis	13	242,513	38.9	26	290,000	0	0
Dallas	14	210,238	24.9	12	528,000	0	0
Newark	15	207,458	54.2	same as New York		0	1
Indianapolis	16	134,320	18.0	16	385,000	0	0
Birmingham	17	126,388	42.0	40	218,000	6,000	3
Cincinnati	18	125,070	27.6	17	379,000	0	0
Oakland	19	124,710	34.5	same as San Francisco		0	1
Jacksonville	20	118,158	22.3	68	143,000	0	0
Kansas City (Mo)	21	112,005	22.1	22	344,000	0	0
Milwaukee	22	105,088	14.7	23	342,000	0	0
Pittsburgh	23	104,904	20.2	10	607,000	52,133	28
Richmond (Va)	24	104,766	42.0	64	156,000	1,700	1
Boston	25	104,707	16.3	6	834,000	3,000	3
Columbus (O)	26	99,627	18.5	27	286,000	5,000	1
San Francisco	27	96,078	13.4	7	769,000	109,855	24
Buffalo	28	94,329	20.4	24	313,000	1,150	2
Gary	29	92,695	52.8	same as Chicago		0	1
Nashville	30	87,851	19.6	30	281,000	0	0
Norfolk	31	87,261	28.3	43	209,000	1,400	3
Louisville	32	86,040	32.8	38	228,000	278	1
Fort Worth	33	78,324	19.9	same as Dallas		same as Dallas	
Miami	34	76,156	22.7	21	349,000	0	0
Dayton	35	74,284	30.5	41	213,000	5,300	2
Charlotte	36	72,972	30.3	42	210,000	15,035	6
Mobile	37	67,356	35.4	60	170,000	10,702	2
Shreveport	38	62,162	34.1	59	171,000	350	1
Jackson	39	61,063	39.7	76	126,000	0	0
Compton (Ca)	40	55,781	71.0	same as Los Angeles		0	0
Tampa	41	54,720	19.7	27	286,000	13,500	4
Jersey City	42	54,595	21.0	same as New York		(5)	
Flint	43	54,237	28.1	62	160,000	8,000	1
Savannah	44	53,111	44.9	below Top 100 Market		below Top 100 Market	
San Diego	45	52,961	7.6	52	193,000	57,591	6
Toledo	46	52,915	13.8	53	191,000	17,650	4
Okla. City	47	50,103	13.7	39	227,000	0	0
San Antonio	48	50,041	7.6	43	209,000	0	0
Rochester	49	49,647	16.8	57	174,000	0	0
East St. Louis	50	48,368	69.1	same as St. Louis		same as St. Louis	

$66,000. The report went on to state:

> . . . Increases in percentage saturation obviously have a most beneficial effect on operations. No new construction is necessary, since the head-end is already in place and the cable network presumably passes all, or almost all, of the homes. The hook-up is nominal and wholly or in part paid for by the subscriber. The actual operating and maintenance costs increase only moderately as additional subscribers are added, total employment on a system of this size being 7 or 8 people. If we assume that after three more years, the saturation percentage has reached 55% or 5,500 subscribers out of the 10,000 potential, the annualized income might look like this:

Revenues (5,500 subscribers at $63 each)[5]	$350,000
Operating Costs	160,000
Operating profit	190,000
Depreciation and amortization[6]	60.000
Interest[7]	23,000
Pre-tax income	107,000
Taxes (obviated by tax loss carry forwards)	
Net income	107,000
Cash flow (normal accounting basis)	167,000

> These figures reflect the economy of more mature systems. By the addition of 2,000 subscribers we have brought the capital investment per subscriber down from about $160 to $109. With 55% of the potential homes subscribing, a system operator might reasonably expect to achieve a pre-tax margin of 30–35%. The cash flow of $167,000 means that at this rate the system cost is being recouped in less than four years' time.
>
> Cable systems can reasonably be expected to become profitable in the third year of operation; the start-up costs, if expensed, produce significant losses in the first two years. The tax loss carry forwards from this early period should run out in about five years. For this reason there is no provision for taxes either at the end of the second or fifth years, although presumably the operations in the sixth year would necessitate payments.

[5] Includes hook-up fee and multi-outlet charges.

[6] Increased by additions to plant.

[7] Reduced by payments on principal.

These projections are fairly reliable for inner city communities though the data on density and cable installation costs will change. An average 200 to 250 homes per mile in the larger cities will create a greatly increased saturation potential.

Since the inner cities and black communities are not yet wired, accurate cost data are not available. There are few variables beyond underground construction, however, that would seriously inflate the estimates made by Drexel Harriman Ripley.

Promotional costs may be higher in the first three years in the inner city. Some of these costs could be absorbed by the city government, community development corporations, model cities programs, civic and social organizations. For example, the city could allocate funds for a promotion campaign against future receipts from franchise fees.

The economic potential of these systems for minority communities goes far beyond the basic revenues derived from operation. A significant number of spin-off business opportunities will flow out of CATV enterprises. The following list is illustrative:

Household supplies
Printing and graphics
Advertising
Vehicle purchase and maintenance
Computer services for billings

Banking services
Channel leasing
Office supplies and equipment
Janitorial services
Local program productions for
  public and private institutions,
  agencies and organizations

Manpower training programs
Construction
Office rental/lease

Business services are expected to generate sizeable revenues for cable systems in the cities within the decade. Some economists estimate that public and business purchases of broad band communication services would range between $40 and $80 billion per year once a national network is completed.

While the development potential is great and the prospects for economic success are favorable, community-owned systems, like all other cable systems, must be able to attract a large number of subscribers to break even or return a profit. Teleprompter, the franchise operator in Harlem, is attracting black subscribers, but does not rely solely on these subscribers for its revenues. Does the opportunity for community ownership provide a sufficient incentive for residents to subscribe? The existing conditions within the community may determine the answer.

---

subscribers – the key to control

As stated earlier, CATV has grown and developed in rural areas and small towns where no television was available or in areas where signal reception was poor.

The major incentives to residents of these rural and out-of-the-way communities to purchase cable services have been (1) better reception and (2) more signals or channels. Similar incentives exist in New York City and a few other major cities where tall buildings and atmospheric conditions cause poor signal quality. However, these conditions are not duplicated in the majority of black population centers. In most central cities there is fair to good signal quality and five to seven broadcast television channels are now available.

Ghetto residents are not likely to subscribe for cable services just to get better signal quality or more channels. Other reasons must be found. The emphasis on community control and development can provide some of the necessary incentives. The few soul radio stations owned by blacks and the total exclusion of blacks from ownership of television stations provide additional motives for local communities to prevent the continuation of these patterns in the cable communications industry. Blacks own none of the more than eight hundred licensed commercial television stations and only about twelve of the three hundred and fifty soul radio stations. Most black communities do not have a local newspaper or magazine produced by and for them.

The strongest incentive for local residents to subscribe to a community owned and controlled cable system may well be the opportunity to combat the insensitive programming of the existing media, the exploitative practices of soul radio stations, and the discriminatory hiring practices of the radio, television and print media enterprises.

The National Advisory Commission on Civil Disorders reported that " . . . . television and newpapers offered black Americans an almost totally white world; and far too often, acted and talked about blacks as if they neither read newspapers nor watched television."

As lawyer Donald K. Hill pointed out, however, " . . . The Commission's Report only touched upon the tremendous impact which the white-culturally-oriented media has on the black community. Although black Americans have the opportunity to fully observe the white world, communication flows in only one direction; blacks never see themselves as they perceive themselves, nor does communication flow from blacks to blacks. . . ."

A more current assessment of the racist practices of radio and television is presented in recent reports in several magazines. The June 24, 1971 issue of *Jet Magazine* outlines the deep-seated and widespread bias in radio programming and in the personnel policies of the regulatory agency responsible for preventing such practices, the Federal Communications Commission.

Jet points out that jazz and other forms of black music are only played on FM or "underground" radio stations and that soul radio stations in most communities operate on a sunrise to sunset schedule and broadcast weaker signals than the white pop music stations. *Jet* also claims that white commercial radio stations conspire with white record companies in a refusal to play black music and that blacks are excluded from top policy-making jobs in the FCC. The FCC has a total of 251 jobs within civil service grades GS-14 to GS-18. Out of this group there is only one black GS-15 and two GS-14s. There has never been a black FCC Commissioner.

Recently *Black Communicator* revealed that the FCC and the broadcast industry have similar patterns of racial bias in employment. In a front-page article headlined "Broadcast Jobs Slim for Minorities," the *Communicator* reported that blacks make up less than 6% of the total broadcast industry work force. Blacks are in only 2% of the manager and official slots and 6% of the professional staff

(FCC Form 395) FEDERAL COMMUNICATIONS COMMISSION

Annual Employment Report: National Summary

SECTION III - FULL-TIME PAID EMPLOYEES (applicable to all respondents) PERCENTAGES+

JOB CATEGORIES[1]	ALL EMPLOYEES[2]			MINORITY GROUP EMPLOYEES				
				TOTAL				
	% Total (Col. 2+3) (1)	% Male (2)	% Female (3)	% Negro (4)	% Oriental (5)	% American Indian[3] (6)	% Spanish Surnamed American (7)	% Total Minority (8)
Officials and managers	100%	90%	10%	2%			3%	5%
Professionals	100%	89%	11%	6%			2%	8%
Technicians	100%	98%	2%	3%			4%	8%
Sales workers	100%	90%	10%	2%			2%	5%
Office and clerical	100%	11%	89%	6%			2%	9%
Craftsmen (Skilled)	100%	96%	4%	8%			2%	11%
Operatives (Semi-skilled)	100%	95%	5%	4%			1%	5%
Laborers (Unskilled)	100%	95%	5%	15%			25%	40%
Service workers	100%	85%	15%	46%			0%	46%
TOTAL	100%	77%	23%	5%			3%	8%
Total employment from previous report (if any)								

[1]Refer to Instructions for explanation of all title functions.

[2]Include "Minority Group Employees" and others. See instruction 6.

[3]In Alaska, include Eskimos and Aleuts with "American Indian."

+ Based on 5% sample of all forms available as of May 27, 1971

* Other minorities represent around .5%, but due to rounding, not reflected in some figures.

category. In the laborer and service categories, blacks are over-represented with percentages of 15% and 46% respectively.

While there are now more minority programs and media professionals on commercial and public television stations than ever before, *Black Enterprise* magazine reports that there is little change in the scope of institutionalized racism in the industry. Author Carol Morton points out:

> Just a bit over 25 years old, television today is a powerhouse of opinion formulation. Ninety-five percent of the nation's households spend more than a quarter of their waking hours in front of a television set. For blacks it's more. Most sets in the black community are turned on at least six and a half hours a day. And, by the time the average child enters the first grade he's already spent more hours in front of a TV set than he will in a classroom earning a college degree.
>
> But, the institution known as television, like most others in this country, has been impermeable as far as black people are concerned. Of the 863 TV stations across the country, not one is black owned or managed. This public resource is concentrated in the hands of a few white individuals.
>
> However, license owners are, by law, required to broadcast in the public interest; to air all sides of controversial issues, putting the interests of the public above their own private aspirations. So, to comply with the law, many stations in areas that serve large black communities do carry black programs. Many of these shows, however, are plagued with miniscule budgets, practically no autonomy, preemption by white station managers, poor time slots, and other problems.
>
> "Everyone making the decisions about what black people will see on TV today is white. That's institutional racism," says Tony Brown, executive producer of educational television's national news show "Black Journal." His program is a classic example of the problem. After the recent airing of a special entitled "Justice?" about the black prisoners and the Angela Davis case, Brown received a forceful letter from a white Las Vegas program and promotion manager who stated that had she had an opportunity to preview the program it would not have been shown.
>
> Station managers on all commercial or educational television channels also decide the budgets of black programs, the time they will be shown, and, in effect, have total control over what goes on the air. Says Charles Hobson, producer of "Like It Is," a black oriented program in New York City shown over WABC-TV, "We have to deal with white rating systems that don't have anything to do with us." The problem of black programs on commercial stations is that they must not offend advertisers. On educational or public television it's a different phenomenon because it is financed with public tax monies allotted by the government. But even though these 202 public TV stations fill some 2,397 viewing hours a month only one hour is devoted to a national black news program—"Black Journal." "In effect," says producer Brown and others, "this means that black people, whose taxes go into public TV, are financing racism."

Blacks have little or no stake in the existing radio and television enterprises in their communities. As a result of widespread challenges to broadcast station license renewals spearheaded by Bill Wright of Black Efforts for Soul in Television (BEST), some long overdue changes are taking place in employment and programming. The broadcast industry is fighting these grass-roots efforts with their lobbyists and lawyers from the largest law firms in the country.

The opportunity for ownership, control, management and programming of CATV systems in the cities can be a powerful incentive to powerless minorities. If they are aware of the economic and social potential of this media for their neighborhood, ghetto residents may be persuaded to subscribe to a community-owned system as a matter of enlightened self-interest. *Five dollars per month is probably the lowest price they can pay to secure a share in the wealth and power of the country.*

Collective ownership and control of systems will undoubtedly enhance the incentive to local residents to subscribe. Shares in the enterprise should be offered to local residents, just as AT&T offers its stock to its employees on a payroll deduction basis. CATV systems could apply a portion of the monthly service charge to the purchase of common stock by the subscribers.

Selling cable to urban residents who already have good reception and multiple channels will not be an easy proposition for whites or blacks. It should not be any more difficult however, for minority entrepreneurs than whites. In fact, it may be easier if the strong desire for community control is recognized and if entrepreneurs are willing to include the residents in the ownership. As Jim Dowdy of the Harlem Commonwealth Council points out:

> There are some enterprises that the whole community must own in order to protect individual black entrepreneuers from the vices of the capitalistic system. Community based organizations and corporations that are responsive to broader community development needs *can* find subscriber support.

---

strategies for developing community systems

Communications experts estimate the cost of wiring each of the major cities will range from $2 million to $20 million. Where will community based corporations get the financing to build and operate CATV systems? Financing will not be simple or easy. On the other hand, it's not impossible. Joint ventures with white or nonresident minority investors are one possibility. Such investors might be banks, national and local church groups, wealthy individuals, insurance companies, savings and loan associations, high income blacks (e.g. athletes and entertainers) and local or national industrial concerns.

Available resources within the black community should not be overlooked. In fact, that's where the initial organizing efforts should start. Many black professionals (doctors, dentists, ministers, lawyers, teachers and businessmen) have relatively high incomes and accumulated savings that can be tapped. The professional class can also provide collateral assets in securing outside financing because of stable employment and high incomes. Black churches and insurance firms are also potential sources of equity capital.

Community development corporations and model cities programs deserve special attention and consideration because of their uniqueness. These groups, operating with both public and private funds, have been established to plan, design and implement community redevelopment and development projects in the inner city ghettoes. One or the other of these entities, and sometimes both, are presently operating in most of the major cities. These groups usually have a working knowledge of city hall politics, the local financial community, the federal funding structure, and the national philanthropic community. They also have the staff and technical expertise to pakage a community proposal.

In short, both the community development corporation and the model cities programs are in a position to act as an effective broker for the community in planning, designing and implementing a program for community control of CATV systems.

There are several other national organizations and programs that can contribute to the community planning and development process. These include:

The National Business League

The National Progress Association for Economic Development

The National Council for Equal Business Opportunity

The Interracial Council for Business Opportunity

The National Banking Association

The Nation of Islam

A complete listing of organizations available to assist in business and community development projects can be obtained from the Office of Minority Business Enterprise, U.S. Department of Commerce, Washington, D.C. 20230

There are several approaches that could substantially reduce the financing burden on local communities. One approach that is practical for large cities like Washington, D.C., is to divide the city into four to six cable districts. If a franchise were awarded for each cable district, four to six cable companies with roughly 30 to 40,000 households could be established. Fifty-five percent market saturation in each district would result in a 15 to 20,000 subscriber system. Under this arrangement, each company would have to raise only $1 to $2 million in financing in lieu of one company attempting to raise $10 to $20 million to build a city-wide system.

Tom Atkins of Boston, one of the most knowledgeable city councilmen in the country on CATV, suggests that municipal governments should "wire-up" the entire city and then take bids for system management and operation. This approach would eliminate the big cost of system construction and place community groups in a highly competitive position for franchise awards. This is an attractive proposition where multiple cable districts are established and the minority community is not fragmented into several predominantly white districts. Blacks, in particular, have been disenfranchised by such gerrymandering in the past.

A common practice in the CATV industry is the 'turn-key' system of construction. Hardware manufacturers have financed, designed and built systems, turning the completed system over to the owner. Some hardware manufacturers prefer to enter into joint ventures for turn-key systems, providing from 30 to 50% of the financing. Care must be exercised, however, to assure that community-owned equity in the system is the controlling interest. The community corporation should also secure an option to buy out joint venture partners on a first sale offer basis.

Community ownership arrangements will certainly influence the source and availability of financing. Several corporate ownership arrangements seem feasible. These are community profit, community nonprofit, a combination of profit and nonprofit, private turn-key ownership, and cooperative.

Community Profit

There are many examples of this type of ownership among enterprises established by community development corporations. A sponsoring organization or group, such as a community development corporation, church, civic or charitable organization could establish the profit-making corporation and maintain voting control over it until the corporation becomes self-sustaining. The corporate charter should provide for this group to relinquish voting control to residents (community shareholders) when the corporation breaks even.

Profits earned by the system could be distributed either in the form of dividends, discounts on service charges, applied to the purchase of additional stock, or paid into a community trust for subsequent allocation to education, health, welfare, recreational programs, or other social needs of the community. Transfer and sale of shares would be restricted to residents of the area served by the CATV system. Private institutional investors, i.e., banks, insurance companies, loan companies, may be better attracted to this corporate arrangement than others.

Community Nonprofit

A nonprofit corporation is neither the *most* nor the *least* ideal arrangement for minorities. Malfeasance, embezzlement, and similar types of capitalistic 'rip-offs' are not eliminated by a nonprofit arrangement. But the peculiarities of nonprofits may enhance local ownership in some cities. A nonprofit corporation may be more attractive to other nonprofits for joint ventures. A consortium comprised of churches, colleges and universities, neighborhood and block associations, and similar groups could be formed. Funds may also be raised through private foundations as well as public bond issues. An advantage of a nonprofit is that it offers outside investors total depreciation of the system to increase their tax write-off. Since the nonprofit does not pay taxes, it doesn't need the tax loss.

The most common rationale for going nonprofit is that it assures that earnings will be allocated for socially useful purposes and that individuals will not get rich while the community remains poor. As outlined in the previous section, however, there are ways to distribute profits on a community-wide basis. Each community should decide between a profit and a nonprofit arrangement based on the priorities and long-term needs of the individual community. In some communities, increasing the per capita income via the distribution of profits will increase savings and purchasing power of the group for other community enterprises. This may be a more effective development program than putting earnings into social programs that cannot regenerate capital.

Combination Community Profit and Nonprofit

Two corporations with equal equity are established, or one group may have a controlling interest. If the profit group is made up of nonresidents or wealthy or high income residents, the nonprofit group comprised of residents and the poor should have the controlling interest. The advantage of this arrangement is that the nonprofit group has very limited responsibility for fund raising, system construction and system management. If profits are earned, however, the nonprofit group would receive a pro rata share of the earnings.

Private Turn-Key Ownership

Two forms of turn-key ownership may be employed. The first method, advocated by CATV consultants and by equipment manufacturers, involves hiring an outside engineering and construction firm to build the entire system. When construction is finished, the community group pays the builder and the key is turned over to the new owner. In this case, the community group (either profit, nonprofit, or joint venture) has to come up with venture capital and financing. Although some equipment manufacturers are willing to arrange, or even participate in financing the system. Under the second form, a municipal government could invite private sources to construct and operate a city-wide CATV system with the stipulation that the system would be turned over to local management groups within five to ten years.

A five to ten year term would enable private investors to recoup their investment plus the profits that most systems realize during that period. Since many operating systems have been sold repeatedly at the end of the depreciation write-off period (roughly ten years) this arrangement may be attractive, especially to hardware manufacturers. Or the city could offer municipal bonds to finance construction and operation.

Two features that community groups should opt for in this second turn-key arrangement are: (1) the training of local residents to assume ownership, management and operational responsibilities prior to system turnover, and (2) an agreement with the municipal authorities concerning eventual local ownership by community residents.

Cooperative

There are approximately 30 to 40 subscriber-owned cooperative CATV systems now in operation. Many of the first operating systems in small towns in Oregon and Pennsylvania are cooperatively-owned. Cooperative, subscriber-owned systems may be ideal for small predominantly minority towns, communities of 10,000 or less, rural communities, and middle-income communities that are so separated from the larger black community geographically that a single-community system is impractical.

Descriptive materials prepared by cooperative systems in Heppner, Oregon and Quincy, California are illustrative of the cooperative model.

QUINCY COMMUNITY TV ASSOCIATION, Inc.

P.O. Box 834 - Telephone 283-2040
81 Bradley Street - Quincy, California
June 10, 1969

The Quincy Community TV Association was founded in 1956 for the purpose of securing adequate television reception for the Quincy area. This brochure has been prepared to provide new and prospective members of the Association with the essential facts concerning the organization. Complete by-laws are available for inspection by those desiring more detailed information.

The Association is a nonprofit corporation governed by a Board of seven members who are elected by the members of the Association for one year terms. The officers of the corporation are elected by the Board of Directors each year and consist of a President, a Vice-President, a Secretary and a Treasurer-Manager. The powers and duties of the Board of Directors and all officers of the corporation are defined in the by-laws of the organization. The subscribers are the actual owners of the system.

Membership in the Association is open to any reputable person, association, corporation, partnership or estate in East Quincy or West Quincy. Each member, regardless of the number of memberships he holds, is entitled to one vote in any general meeting of the Association, which are held annually in July.

The physical plant of the system consists of about twenty-six miles of aluminum shielded maintrunk line in Quincy and East Quincy. The antennas are located at an elevation of 6,994′ on top of West Claremont Peak. Twenty-Five Thousand Feet of buried armored cable brings the TV signals to Quincy. The total cost of the system is about $180,000.

Prior to December 1968, membership in the association cost $125.00. For members who paid this amount the monthly service charge is presently $5.00 per month.

Effective December 1968, new members are connected for a fee of only $25.00. For these subscribers the monthly service charge is $6.00 per month. Eventually, when the finances of the association permit, it is hoped to place all subscribers on an equal basis.

A $1.00 discount is given anyone paying six months service charge in advance. Service will be suspended to any member two months in arrears on his monthly service charge. Reinstatement is then available upon payment of a $10.00 fee. A subscriber may move his hookup any place in the Quincy area at a cost of $10.00.

Secondary services available are FM Radio splitters at a cost of $7.00 for materials. Also a subscriber may have a second TV outlet in his home for a cost of $12.00. There is no extra monthly charge for these services.

Persons desiring FM Radio only are hooked up at a cost of $25.00. The service charge for FM Radio only is $1.00 per month.

FACTS ABOUT–
Heppner TV Inc.
"TELEVISION BY CABLE"
AFFILIATED WITH
Pacific Northwest Community
TV Association, Incorporated

Heppner TV Inc.
289 North Main P.O. Box 587
Heppner, Oregon
Phone 676-9663

Formed In 1955

Heppner T.V., Inc. is a cooperative formed in 1955 to obtain television pictures for the City of Heppner.

Due to the terrain and the distance to the closest television station it was and is impossible to obtain any free air signals within the city. For this reason it was decided to construct a cable system to bring the signals to town from a hill northwest of the City Hall.

In order to make this possible, membership certificates in the cooperative were set at $135.00 to raise the money to construct the system. These certificates belong to the individual and are saleable if the person so desires.

Channels Listed

In the beginning there was only one channel available on the system and in subsequent years this was raised to three and then to four plus one channel of FM. The channels now available on the system are:

KATU Channel 2 Portland
  Heppner TV Channel 2
KPFM Music
  Heppner TV Channel 3
KGW Channel 8 Portland
  Heppner TV Channel 4
KOAP Channel 10 Portland
  Heppner TV Channel 5
KPTV Channel 12 Portland
  Heppner TV Channel 5
KOIN Channel 6 Portland
  Heppner TV Channel 6

Microwave System

All the Television Channels are now microwaved to the hill Northwest of town from a receiving and transmitting site at Goodnoe Hills, Wash. This microwave link is owned by a common carrier company which leases the service to Heppner T.V., Inc.

At the present time channel 5, Heppner, carries both KOAP Educational channel 10, Portland, and KPTV, the Independent Station Channel 12, Portland. At the time of constructing the microwave system the Federal Communications Commission stated that KOAP Channel 10 had to be carried on the system at all times. Due to the high cost of microwave equipment the company could only afford to place four channels on the air, which is the reason you receive channel 10 at all times it is on the air, and when it is shut down, channel 12 is automatically turned on for the remainder of the broadcast day.

You also can receive some FM radio stations by connecting the cable to an FM radio. These stations come from Portland or Kennewick, Wash., direct to Heppner TV's receiving site and due to the distance involved some fading is experienced, particularly during the dry summer months and at sundown.

Rates and Hook-up Charges

Memberships

Membership in Heppner TV, Inc.—$135.00.

This includes the installation fee and all rights as a member of the cooperative.

This membership may be purchased at the company office or from another member if he so desires to sell for his asking price.

The membership fee may be financed at the company office for $15.00 down with payments to the bank of $10.85 per month until the balance of $120.00 is paid.

Normal service charge to members is $4.00 per month. Temporary service (as explained below) requires a $5.00 monthly service charge.

If the membership is purchased from another member there will be an installation charge of $7.50 if the cable is already in the residence, and $15.00 if a new installation has to be made.

If the member moves to any other town in the United States affiliated with the Pacific Northwest Community Television Association, Inc., the connection is transferrable which in most cases results in a lower monthly service charge. By the same token, connections are transferrable into Heppner with a hook-up fee of $15.00 if the system is a member of any one of the various associations throughout the country affiliated with the PNCTA.

Temporary Service

Heppner TV, Inc. has a plan for a temporary resident which is good for six months only from the date of purchase.

Hook-up Charge	$15.00
Deposit	10.00
1 Month in advance	5.00
Total	$30.00

At the end of six months the Hook-up Charge may be used as a down payment on the Regular Membership Certificate. As noted, the monthly service charge under this plan is $5.00.

Additional Sets & FM

If you wish an additional connection in your house for either another television set or an FM radio the cost varies according to the cost of material and labor involved. The average cost of an extra outlet is $20.00. Heppner TV, Inc. does *not* make any additional monthly charge for this service.

municipal ownership— beware!

Another ownership arrangement that now exists in eight or nine cities is municipal ownership. Municipal ownership may or may not insure the expenditure of revenues on local programming or other social programs that will promote community development. Revenues may be placed in the general city fund and used for city-wide projects. Municipal ownership could be financed by low-interest public bonds; however, the current financial crisis of the cities may make this approach unpopular.

Municipal ownership could take the corporate form of the public broadcasting systems that now exist in several states, or a corporate form similar to a Port Authority. A municipality could also contract with a hardware manufacturer to build the system, leasing it to a profit or nonprofit group on a long-term basis.

Even in cities where blacks are a majority of the population, municipal ownership could be a shaky proposition. Municipal ownership sets a swarm of legal and procedural requirements in motion that could hamper community control, ownership and participation. It is not unusual for local governments to avoid actions that clearly enhance black development on the basis of 'reverse discrimination' claims. Minorities seeking guarantees that a municipally-owned system would provide for community participation in the planning, programming, management, employment and distribution of earnings would find themselves competing with an array of interest groups within the total community. Although white civic, social and political groups have not been disenfranchised and exploited the way blacks have been, local elected officials are unlikely to make this important distinction.

Local governments, power blocs and the white electorate will probably resist minority ownership and control of CATV systems under any circumstances. Municipal ownership could be a convenient mechanism for maintaining white control. Blacks are woefully under-represented in the policy-making structure of existing municipally-controlled agencies and corporations such as public housing, port, and airport authorities. Unless there are unusual legal proscriptions that favor and protect minority control of community systems, similar conditions of exclusion and under-representation are likely to emerge in the policy-making bodies of municipally-owned systems.

---

the spy in your bedroom

Cable television can deliver economic, political, and communications power to urban minority groups that take action. This potential for power is only part of the urgent need for immediate action and involvement by blacks. Cable systems can also cause great harm. Some of the harmful aspects conjure up scenes from science fiction. In a recent issue of *The Nation*, there is a comprehensive discussion of CATV and broad band communications titled "The Wired Nation." Ralph Lee Smith wrote:

> . . . It cannot be assumed that all the social effects of the cable will be good. For example, the exodus of

the middle and upper-middle class from the cities is expected to continue during the 1970's, and the stratification of society along geographic-economic lines will thereby be increased. At the same time, the cable will make it less and less necessary for the more affluent population of the suburbs to enter the city, either for work or recreation. Lack of concern and alienation could easily deepen, with effects that could cancel the benefits of community expression that the cable will bring to inner-city neighborhoods. At the very least, such dangerous possibilities must be foreseen, and the educational potential of the cable itself must be strongly marshalled to meet them. The bland treatment of this issue . . . is chilling.

There are other serious problems. Police surveillance by cable, and the compiling of financial, credit, and other personal information about individuals in computer banks, raise unprecedented issues of civil liberties and privacy. Privacy problems are also involved in the transmission of mail by cable. And in the creation of Facsimile Data Services . . . the question is, who will decide what data is to be included in these services, and what is to be left out? Unless the issues involved in these future uses of the cable are understood and faced, 1984 could easily come well in advance of George Orwell's prediction.

Mr. Smith's portrayal of the dangers is informative but it does not go far enough.

If the "wired nation" becomes a reality, clandestine electronic surveillance will be greatly simplified. Experiments involving the use of video cameras mounted on poles in the business district of several cities, including downtown Washington, D.C., are already underway. By the use of a central television monitor, one policeman can patrol the entire area and alert prowl cars by radio of suspicious activity. This system of electronic snooping is easily transferred to residential areas, apartment buildings and businesses without the knowledge or cooperation of the occupants or owners.

Sophisticated electronic sensors and surveillance devices are in regular use in Vietnam. Much of this technology has already been modified and adapted for domestic security systems. If these devices are coupled with cable technology and satellite systems, an electronic Frankenstein could be created to maintain law and order in the ghettoes. Widespread construction of cable systems will greatly expand and accelerate this law and order activity.

According to the *Washington Star*, (June 10, 1971), a nationwide police-operated computerized intelligence data network is less than a year away. Once communities are wired up, electronic snitching by paid informers could be fed directly into the national network. It would be impossible to identify the informer.

An equal threat to minority communities is the prospect of even more depersonalized social institutions via cable. The toehold gain in decentralizing control of schools and public agencies could be wiped out. Cable television has already produced classrooms without teachers. Children will be taught at home or other places without a teacher being physically present. Case workers can service their clients without direct contact. Birth control information can be transmitted directly into the home. "Julia," "Sesame Street," and "Captain Kangaroo" have demonstrated the impact that television has on children as a teaching and socialization instrument for white middle-class values.

Where is the accountability in such a depersonalized system? How to you organize to protect the interests of the community? An army of actors, writers, producers and directors, each practicing his craft, make up a "Captain Kangaroo" or a "Sesame Street." What happens if public agencies and institutions adopt similar methods for the delivery of services and information?

Community control of CATV systems could be the most effective way to bring an end to the communications void that has existed far too long in minority communities. It is the *only* means for assuring that CATV does not become the modern, electronic slave master that perpetuates economic exploitation, political repression, and cultural genocide against minorities in America.

BIBLIOGRAPHY

"An Industry Report on Community Antenna Television (CATV)," Drexel Harriman Ripley (now Drexel Firestone), October 15, 1968.

"Broadcast Jobs Slim for Minorities," *Black Communicator*, June 1971.

"Cable Television Cash Flow Projection," Communications Publishing Corporation, undated.

"Dialing in Black Radio," *Black Enterprise*, November 1970.

"Fight Bias Against Famous Blacks," *Jet*, June 24, 1971.

"Some Other Choices Besides Off/On," *Black Enterprise*, August 1971.

"Smoke Screen Cloaks Cable TV Czars," *The Christian Science Monitor*, August 7, 1971.

"The Wired Nation," *The Nation*, May 18, 1970.

Toffler, Alvin, *Future Shock*, New York, Bantam Books, 1971.

CHAPTER 3

cablecasting: local origination for cable television

by ted ledbetter

introduction

After 20 years of carrying just broadcast television signals into rural and suburban areas, cable television has begun to offer an alternative to mass audience, network produced, advertiser supported, commercial television. Some systems are simply continuing to fill extra channel space with automated time, weather and news service. Others have constructed and equipped cablecasting studios capable of continuous live production. A few systems even have remote units for cablecasting local high school sports, city council meetings and special events.

The demand for cable access is expected to grow rapidly in the next ten years as cable systems are built in the big cities. The rate at which this demand grows, and the degree to which cable meets the needs and desires of people will depend heavily on the content, management, technology and financing of cable origination. This chapter presents some of the issues and problems of local origination. The information is based on industry data, foundation-sponsored research and the actual experience of SDC–TV, cable Channel 7, in San Diego, California.

cablecasting objectives

Minority groups denied access to the existing mass media should have a prime interest in the potential of cablecasting or local origination programming. Cablecasting can provide an influential voice and generate long-term, stable community development through low cost, locally produced programming *by* and *for* minority groups. But cablecasting, despite (or maybe because of) its apparent low cost, has a long uphill battle to fight before it can attract even the basic local audience necessary for survival.

Broadcast television has conditioned most viewers to respond to only the well-known, network TV performer, whether the performer be a singer on a variety show or a razor blade in a commercial.

A cablecaster trying to match this kind of strong competition would be waging a losing fight. But if cablecasting attacks the weaknesses of broadcast television, it may produce the kind of media outlet that eventually dominates the home TV receiver. Until that day comes, there are several basic, general guidelines that should be followed:

1. *Keep It Local*—Cablecasting can attract local viewers who cannot otherwise participate in local events.
2. *Use The Best Equipment Available*—Viewers will expect "broadcast quality" including color transmission.
3. *Accept Expert Advice*—Ask questions, listen and seek help when you need it.
4. *Share The Experience*—Some of the most creative and successful programming and technical ideas will come from the least anticipated sources including artists, children, students and viewers.
5. *Look Before You Leap*—Check with cable system operators, check the local CATV franchise and know the regulations.*

*See Chapter 4 following, "Municipal Regulations" and Section III, "FCC Regulations, Cablecasting Regulations, and New York Public Access Rules."

regulations

At the present time, there are two basic types of regulations which govern the use of cable television for local program origination. The first type is explicit and consists of the Federal Communications Commission (FCC) rules 74.1101–74.1121, which include (1) requirements for equal time, fairness, sponsorship identification and (2) prohibitions against lotteries, political censorship, payola, certain forms of pay-TV and commercial advertising except during natural breaks. The second type of regulation is not explicitly applicable to cablecasting, but in fact, does govern it. This type includes the cigarette commercial ban for broadcasting and the general libel, slander, incitement, obscenity and pornography laws. Although most of these laws do not mention cablecasting, their effect is real because few cable operators would be willing to risk any anti-cable reaction at this critical stage of cable's growth.

The following briefly describes the fundamentals of cablecast regulations.

Equal time

As in broadcasting, an opposing candidate for the same office in the same primary or general election must be granted equal time for reply upon request. Equal time means time of equal value, but does not necessarily mean the same time of day or on the same program. (NOTE: The rules do not require cablecasters to initially grant any time for political use.)

Fairness Doctrine

Cablecasters are required to afford a reasonable opportunity for opposing views on issues of public importance. They are also required to *solicit* responses to personal attacks except when made on foreigners, or by legally qualified candidates, or during newscasts, interviews or on-the-spot news coverage.

Sponsorship Identification

Just like it sounds.

Lotteries

No lotteries (consideration/chance/prize) allowed.

Political Censorship

No censorship permitted of material cablecast by any legally qualified candidate for public office.

Payola

Indirect "sponsorship" must be identified.

Pay-TV

Cablecasters cannot charge a per program or per channel fee for feature films which have been in general release more than two years preceding the cablecast. Also, these charges cannot be made for sports events which have been broadcast within the preceding two years. There are a few allowable exceptions to these rules.

Commercial Advertising

Cablecast advertising is permitted only during natural breaks, which are (1) between programs, (2) during time-outs in sports, (3) between acts of plays, during recess of city council meeting, and (4) during movie intermission "present at the time of theater distribution."

Now that you know the basic rules, you should find out how to get on the cable with your program.

---

channel access

There are three basic arrangements that a would-be cablecaster can make with the cable company in order to get access to a channel.

1. *Pay For It*. Most cable systems with extra channel space will welcome an arrangement where you supply videotaped programming and pay a flat fee or a percent of your revenues (if you sell advertising). Avoid this kind of deal unless you are certain you can still cover costs.
2. *Free Exchange*. Several existing systems (e.g. Teleprompter and Sterling in New York) are required to provide public access channels as a condition of their franchise. Assuming that other producers haven't filled up those channels, then you can get on at no charge simply by meeting the terms or rules. (For New York's rules, see section III.)
3. *Sell It*. If you are able to produce programming on which the cable company can sell commercials, they must pay you to do so. Cable systems are not eager to buy such programming unless they are confident of viewers and sponsors. This usually means sports and movies (or foreign language programs in certain areas).

Since many of the newer systems in large cities will require public access channels, we will assume that much of the programming will be supplied on a Free Exchange basis, otherwise referred to as barter. Several large advertisers, with good bartering experience on broadcast television, are trying the same method with cable. It works like this:

1. Producer makes program on videotape and puts in his own commercials.
2. Producer offers taped program with commercials to cable system in return for free time.
3. Cable system plays complete tape on system.

There is no exchange of money, but both parties gain. The producer is paid by the advertiser(s) and the cable system signs up new subscribers attracted by the special programming. As mentioned above, this arrangement usually works best with known-audience programs, like "boxing from Las Vegas" or "Roller Derby."

In general, getting access to the cable is like driving your father's car for the first time. You may have to "pay" him by taking out the trash from the basement, but once he trusts you with the car and finds out how useful it can be for him, he may wind up paying you to drive the car and run errands for him.

In other words, evaluate the strengths and weaknesses of your program, negotiate with your strengths and don't be afraid to renegotiate as the cable operator begins to see what you can do for him.

Important Note:

Some cable systems import distant signals but blank out the portions of those distant signals which duplicate local off-the-air programs. If you can get access to that blanked-out time, do it, because you will have a ready-made audience watching the distant signal when you start.

facilities and equipment

The facilities and equipment required for cablecasting vary widely. You will have to determine what your needs are, what your resources are, and to somehow match the two so that you produce quality programming.

Basic Facilities

Basic facilities for cablecasting consist of a soundproof, air-conditioned room with a good supply of electrical outlets. Although everything can be done in a single room, it's helpful if a smaller, separate room is available as a control room. Cameras, lights, microphones and props are in the main studio (15' x 25' minimum) and the other equipment is isolated in the control room where the noise of the videotape recorder and the director's voice is not picked up by the microphones. (See Figure 1.)

Lighting

Good lighting is one of the most important considerations for good video pictures. As a minimum, floodlamps purchased at a local hardware or department store can be hung from pipes supported overhead. For color productions, professional color spotlights should be used. If there is any doubt about what type to use, try to rent several types and actually test them in the studio.

Props

Basic props include two comfortable chairs similar to those used on TV "talk" shows, a sofa, a couple of tables and some sort of backdrop. Off-white or light colored drapes can be used or, perhaps a textured wall made of egg cartons. Use your imagination. Just remember one rule: don't make the set so interesting that it distracts from the subject.

Required Equipment

A basic TV studio for videotaping for cablecasting consists of two TV cameras, a switcher, TV monitor, videotape recorder, lights, microphones and audio mixer. The switcher is used to switch from one camera to another. The monitor shows what is being recorded on tape, and the audio mixer controls the inputs from the microphone. Both the video and audio are recorded on the same tape at the same time by the videotape recorder.

More élaborate studios would contain more and better cameras and recorders (e.g. color), a monitor for each camera and a film chain, which would allow movies and 35 mm slides to be part of the program. Several types of typical equipment systems are listed below.

Basic Equipment for Cablecasting*

	Black and White (B/W)		Color		
Equipment	Basic	Average	Basic	Average	Full
Figure		II	III	IV	V, VI
Cameras	2	2	2	2	3
Tripods	x	x	x	x	x
dollies		x	x	x	x
zoom lens	x	x	x	x	x
headsets	x	x	x	x	x
Control					
switcher	x	x	x	x	x
special effects	x	x	x	x	x
Film Chain		x		x	x
Audiotape player		x		x	x
Monitors, Video	1	4	3	4	5
Waveform Monitor				x	x
Microphones	2	3	2	3	4
Mixer, audio	x	x	x	x	x
Audio Monitor		x	x	x	x
Video Recorders	1	1	1	2	2
tape size	1/2"	1"	1"	1"	1"
Lights	x	x	x	x	x
Approximate Cost	$10,000	$20,000	$35,000	$50,000	$80,000

*The numbers in each column indicate the minimum quantity needed of each item; an "x" indicates that the item is required, but will vary according to scope of operation, size of budget, etc.

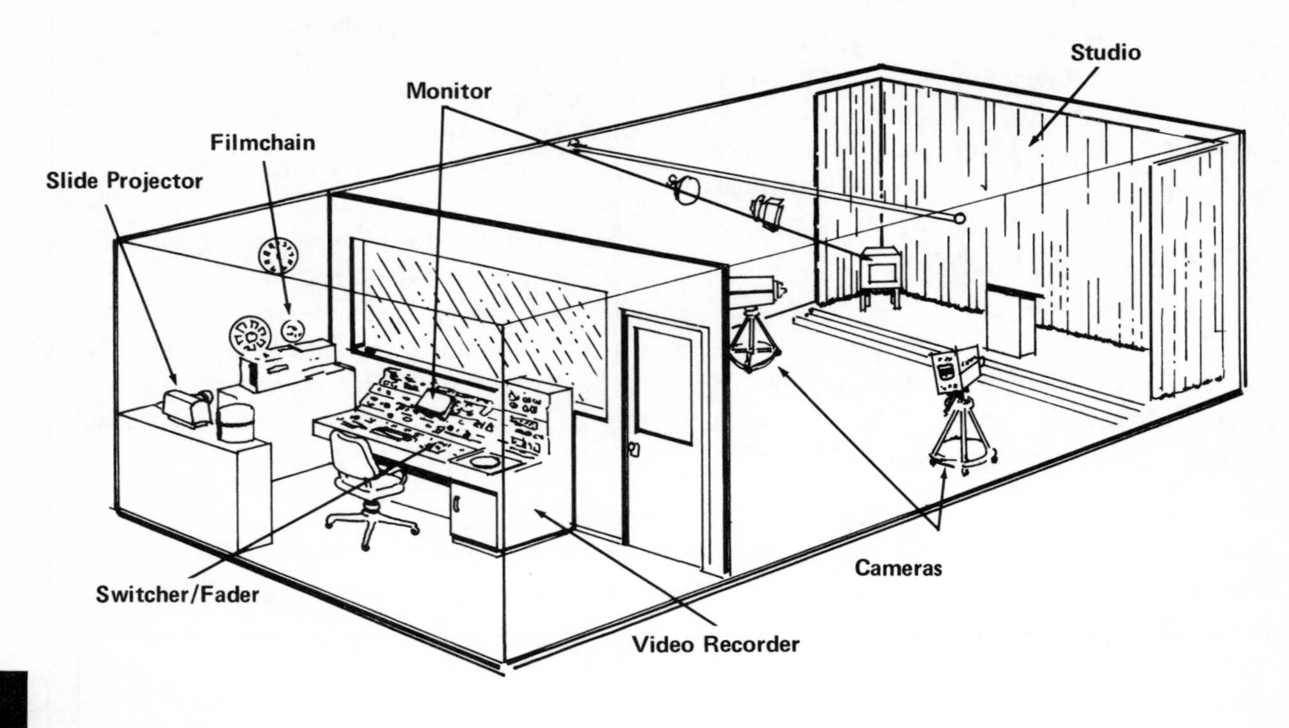

Cablecasting Facility
Figure I

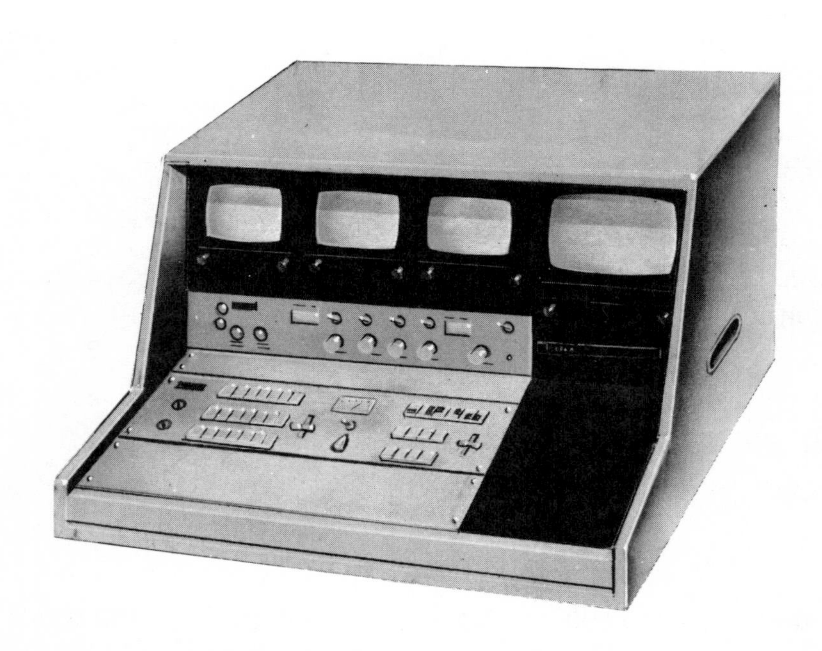

Basic Origination Control Console
Figure II

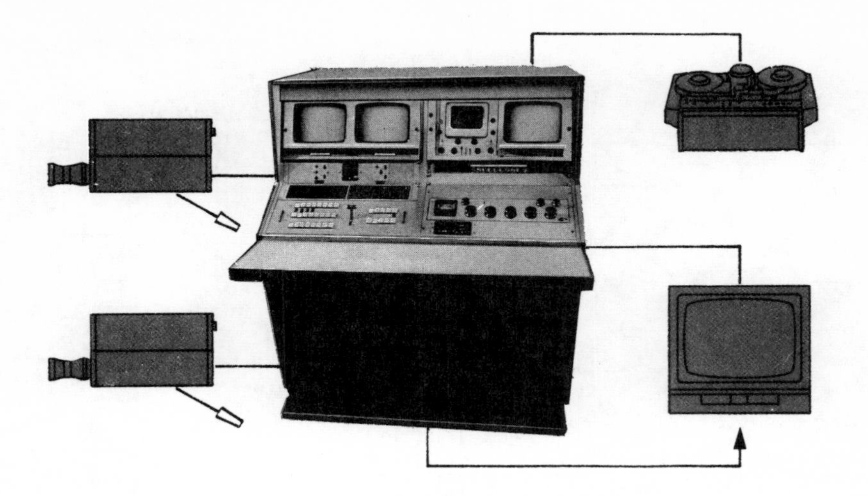

Color Origination Control Console
Figure III

Different lenses will modify these costs as well as various editing features of the recorder. Microphone prices vary also, but the most useful will be $40–50 lavalier mikes that can be hand-held or looped around one's neck.

Videotape Interchangeability

A major consideration in selecting equipment is interchangeability and compatibility. First, all of the equipment must work together. However, not all cameras are alike. Nor is the videotape necessarily interchangeable.

In the ½ inch tape size, all tapes made on the Sony AV series, Panasonic NV-3000 series or any other machine designed to the EIAJ-1 standard are interchangeable. But the Sony AV series is not compatible with the older Sony CV series; nor is the current Panasonic NV-3000 compatible with older Panasonic tapes. In the 1 inch tape size, the only compatibility is between machines made by the same manufacturer but sold by different companies. For example, the 1 inch videotape recorders sold by Bell & Howell, RCA and International Video Corp. (IVC) are all made by IVC. Therefore, tapes made on these machines are interchangeable. Ampex is the largest manufacturer of 1 inch recorders, but it is difficult to find compatibility even between different Ampex machines of the *same* model.

Other Parts

There should always be spare lamps, fuses, cables and videotape. Since videotape is expensive, it should be a major budget item, not an afterthought. In small quantities, new ½ inch video tape costs $40 per hour and 1 inch tape costs $50 per hour. It can, of course, be re-used (just like audio tape), but not more than 20 or 30 times (for B/W) before drop-out spots will begin to mess up the picture. Also, know your local dealer so that you will be able to borrow a spare recorder or camera while yours is being repaired. Other small items needed include, head cleaning fluid, audio tapes for background music, posters for titles, and necessities like telephones and automobiles.

Finally, you must have a way of getting the videotaped program to the CATV system head-end in order for it to be shown over the cable. The least expensive method is to hand carry the tape by car or bicycle and have it fed into the cable system by the systems' video recorder and modulator. But, if the original recorder and the one at the head-end are incompatible, it will be necessary to purchase an extra unit or carry the original one

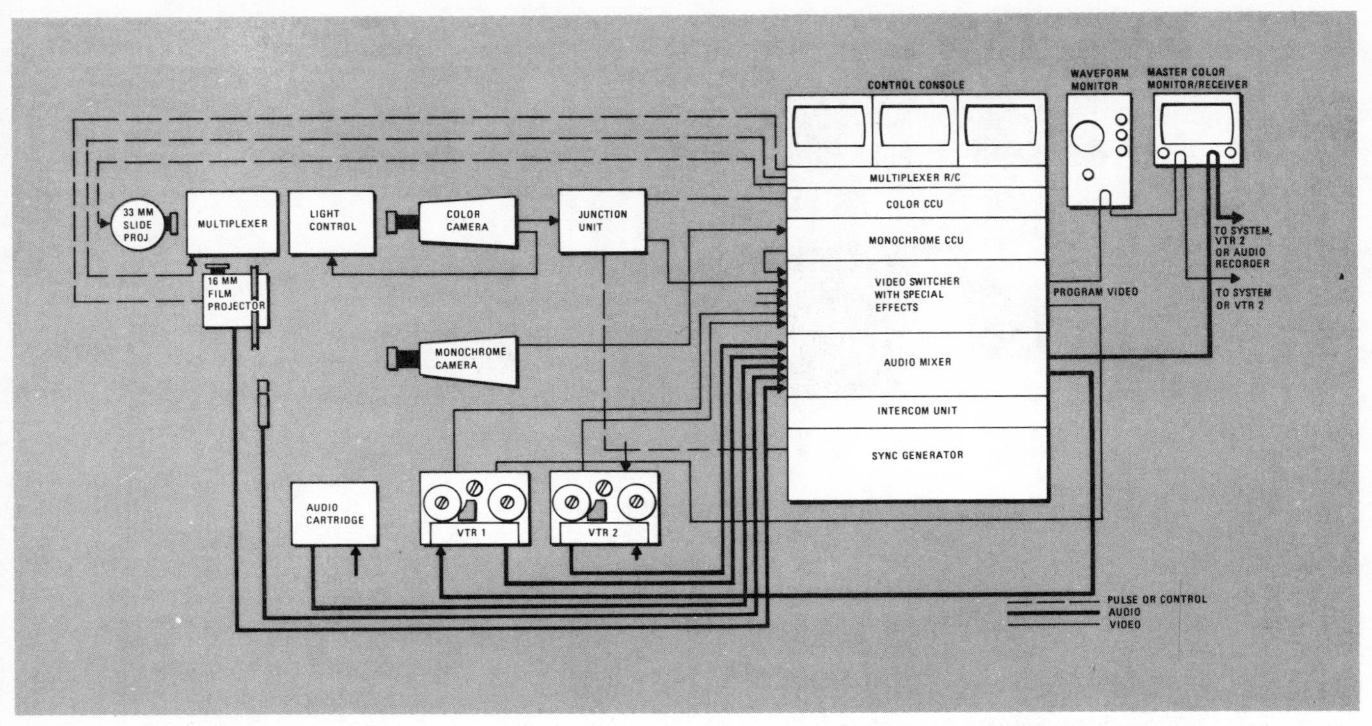

Cablecasting System for Live Black-and-White Production and Color Film and Slide Multiplexing
Figure IV

Full Cablecasting System
Figure V

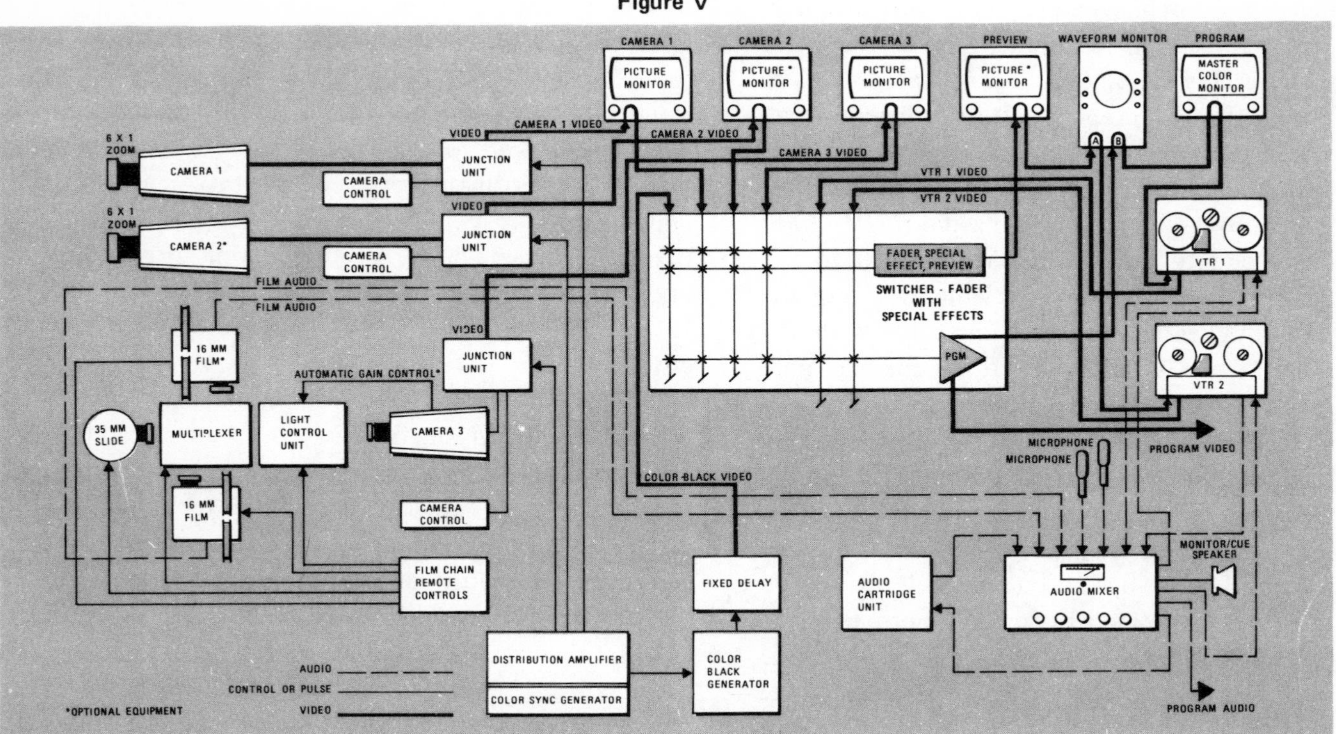

to the head-end each time a program is played.

An alternative is to install a special cable or microwave link from the studio to the head-end. Hopefully, all new cable systems will have a two-way capability so that studios can originate from anywhere along a trunk line.

That's it as far as equipment goes. If it is impossible to purchase and set up an independent studio, check with a local school or university. Or maybe even the cable system itself. They may make it available at a low cost, or even free.

Recommendation

The word quality is subjective, but in terms of local origination CATV, it simply means that if a lot of people on the system have color TV sets, and if most of the other programming is in color, then you must use color cameras, color video recorders and color lighting to effectively compete for your audience. A second example is the difference between ½ inch and 1 inch videotape. First generation unedited masters on ½ inch videotape are acceptable for most newer cable systems. But if you plan to do any editing, use 1 inch videotape equipment, because most ½ inch tape currently available cannot be edited without producing an annoying vertical roll or horizontal tear in the picture.

The choice between black and white or color and between ½ inch or 1 inch tape equipment is primarily dependent upon your resources. Color equipment costs at least three times as much as black and white, and 1 inch equipment costs twice the equivalent in ½ inch. If you can afford to buy, borrow or otherwise acquire color equipment with 1 inch videotape, do it.

Color Cablecasting Facilities
Figure VI

personnel and other expenses

Any of the studio systems listed earlier need a basic three person crew—one for each of two cameras and one operating the control room equipment, including switcher console, videotape recorder, microphone mixer and film chain. Since independent cablecasting is likely to be a nonprofit proposition for the near future, experimentation with the number and types of personnel should be tried.

Other people can be used as producers and studio directors, actors, hosts and technicians. Always have access to someone who can make minor adjustments and repairs to the equipment.

The nonproduction side of the operation will need a manager (for time, money, etc.) and a salesman (if commercial time is sold). These functions can be combined into one person, but it will limit his effectiveness.

Total Origination Costs

The FCC ruled in 1970 that all CATV systems with more than 3,500 subscribers must originate some local programming by April 1, 1971. This rule was overturned in May by the Courts. But on April 1, many CATV systems filed waivers with the FCC to show that they could not afford to originate.

The following examples of projected origination costs (A&B) and recent actual origination costs (C&D) are from four waiver applications.

exhibit A

FIRST YEAR CATV ORIGINATION COSTS

The following projections are based on minimum color origination equipment, which includes local studio and remote studio. The two live color cameras assigned to the remote studio would also serve as live origination equipment for the main studio. Projections are for four hours of programming per day, five days per week—one-half of which would be produced locally, and one-half purchased from outside sources.

Interest on $82,000.00 Equipment Investment	$ 8,280.00
Staff:	
1 Director	9,600.00
2 Cameramen	8,000.00
1 Tech. Director	8,400.00
On-camera Talent	10,400.00
Programs:	
Purchase (Outside sources)	10,800.00
Local Production	12,600.00
Power & Light	1,440.00
Telephone	900.00
Transportation	1,200.00
Secretarial	5,200.00
Accounting	2,600.00
	$ 79,340.00
G & A (@ 19%)	15,075.00
Total 1st Year Orig. Costs	$ 94,415.00
Revenue Expected:	
18 Commercials per day @ $10.00 each	$ 46,800.00
Cost of Sales & Commercial production	23,400.00
Revenue after Commercial costs	$ 23,400.00
Profit (Loss) on Origination	($-71,015.00)

exhibit B

ESTIMATE

OPERATING COSTS FOR LOCAL ORIGINATION OF 20 HOURS PROGRAMMING PER WEEK

PAYROLL	
1 Producer/Director/Technician	12,000
2 Operators – @ $3.50 hour	14,000
1 Talent	12,000
Total Payroll	$38,560
OPERATING SUPPLIES	
Tape	720
Tubes & Lights	600
Repairs	1,000
Total Supplies	$ 2,320
Rent—1,000 square feet of studio and storage space	2,500
Transportation—vehicle	1,200
Sales Expense	4,800
Miscellaneous	1,000
Total Operating Expenses	$50,380
Depreciation ($50,309 over 6 years)	$ 8,385

ESTIMATE

Capital Costs

Color Local Origination Facility for CATV

Quantity	Description	Unit Price	Total Cost
2	Color Vidicon Cameras	$10,000	$20,000
2	Zoom Lens	1,595	3,190
2	Pan Head & Tripod w/Dolly	512	1,024
1	Sync Generator	1,500	1,500
2	9" Monitors	455	910
1	Switcher	2,995	2,995
1	529 Waveform Monitor	1,100	1,100
1	14" Color Monitor	1,925	1,925
3	Intercom Headsets	40	120
1	Audio Mixer	595	595
1	Audio Monitor Amplifier & Speaker	250	250
2	Microphones	100	200
2	Color Videotape Recorders		13,000
500 Ft.	Camera Cable		1,000
	Cabinets & Racks		1,000
	Lighting Equipment		500
	Studio Installation		1,000
		Total Capital Cost	$50,309

exhibit C

ACTUAL

Income Statement—July 1968
New England Cablecasting

	July 1968		Prior Month	Twelve Months Year to Date		
	Budget	Actual	Actual	Budget	Actual	Prior Year Actual
INCOME:	500	—	—	3,000	1,163	—
COST OF OPERATIONS:						
Salary-Director	800	850	850	9,600	9,900	—
Salary-Technical	350	300	279	2,950	2,579	—
Repair & maintenance	50	156	122	660	921	—
Automotive expense	35	43	42	510	773	—
Automotive rent	40	42	42	240	252	—
Power	30	101	86	180	268	—
Talent	575	485	496	5,600	6,257	—
Tape, film, cards	75	44	27	450	103	—
Production supplies	20	2	2	420	344	—
Total	1,975	2,023	1,946	20,610	21,397	—
SELLING, GENERAL & ADM.:						
Salary-office	350	311	216	2,100	1,571	—
Rent-office, etc.	280	280	280	1,680	1,658	—
Office supplies & expenses	50	39	35	480	418	—
Telephone	50	67	55	420	474	—
Advertising-promotion	250	100	100	1,500	337	—
Postage	50	6	—	300	39	—
Travel & ent.	10	—	21	60	1,045	—
Payroll taxes	90	81	73	625	533	—
Insurance-general	25	—	—	150	260	—
Insurance-payroll ben.	25	20	20	150	121	—
Total	1,180	904	800	7,465	6,478	—
Total expense	3,155	2,927	2,746	28,075	27,875	—
CABLECASTING	2,655	2,927	2,746	25,075	26,712	—
CAPITAL EXPENDITURES	$ —	—	320	15,200	15,200	—

exhibit D

ACTUAL

Income Statement—July 1969
New England Cablecasting

	July 1969		Prior Month	Twelve Months Year to Date		
	Budget	Actual	Actual	Budget	Actual	Prior Year Actual
INCOME:	—	—	—	1,800	1,571	1,163
COST OF OPERATIONS:						
Salary-Director	710	537	537	9,360	9,321	9,900
Salary-technical	1,640	1,320	1,397	11,940	9,208	2,579
Repair & maintenance	75	325	39	900	1,536	921
Automotive expense	60	38	17	815	355	773
Power	90	86	109	1,080	923	268
Talent	700	105	204	8,040	5,190	6,257
Tape, film, cards	50	44	—	2,950	3,175	103
Production supplies	60	17	64	570	653	344
Total	3,465	2,475	2,370	35,975	30,397	21,397
SELLING, GENERAL & ADM.:						
Salary-office	255	—	202	3,060	2,695	1,571
Rent-office, etc.	350	350	350	4,100	3,780	1,680
Office supplies & expenses	50	106	57	600	1,030	418
Telephone	60	98	101	720	1,000	474
Advertising-promotion	430	—	119	4,770	1,946	337
Postage	—	68	45	—	505	39
Travel & ent.	20	—	57	380	592	1,045
Payroll taxes	165	101	139	1,470	1,399	533
Insurance-general	30	28	28	300	545	260
Insurance-payroll ben.	75	40	40	630	326	121
Total	1,435	791	1,138	16,030	13,818	6,478
Total expense	4,900	3,266	3,508	52,005	44,215	27,875
CABLECASTING	4,900	(3,266)	(3,508)	50,205	(42,644)	26,712
CAPITAL EXPENDITURES	$ —	428	3,413	8,465	9,984	15,200

four case studies

Four major case studies in local origination have been conducted within the past two years to provide analytical information to policy makers involved in developing cable television regulations and to prospective program producers. The first three cases were described by N. E. Feldman in *Cable Television: Opportunities and Problems in Local Program Origination*, September 1970, (see Section III, Research and Demonstration Projects, Rand Corp.) Feldman's research was performed under a Ford Foundation grant to the Rand Corporation as the third in a series on cable television. Feldman reported on the status of local origination in Canada; Dale City, Virginia; and Lakewood, Ohio.

The fourth recent case study was conducted in San Diego, California by this author at the request of The Urban Institute.

Canada

Canada has several large, profitable cable systems. In Montreal, there are more than 100,000 subscribers and one Montreal cable system has been originating programming for about ten years. Feldman's research indicates that local origination in Canadian cable systems is successful, although mostly uncontroversial. There are several reasons for the success, but the primary reason appears to be the financial support given to local origination by the cable systems, which are viable because of profits from the large subscriber base attracted by the importation of distant signals.

Dale City, Virginia

Dale City Cable TV, in an isolated suburb of Washington, D.C., attracted subscribers because of poor off-the-air reception of broadcast signals. But even though the cable operation was practicable, local origination failed—primarily for financial reasons. Dale City Television, which operated the community channel, refused advertising and tried to operate with volunteers. Also, the monthly cable service fee to subscribers was only $1.50 (the U.S. average is $5.25), and little financial support was available from that source.

Lakewood, Ohio

Lakewood is within the strong local TV signal area of Cleveland, which meant that cable transmission of off-the-air signals provided little incentive for people to subscribe. The owner tried to use local origination as an incentive, but could not meet the competition of off-the-air broadcasting. Because of these problems, the entire cable system failed.

San Diego, California

San Diego has a large viable cable system which imports distant signals from Los Angeles and feeds them along with local signals to more than 40,000 subscribers. San Diego has an independently operated local origination channel. In order to survive, the independent channel (SDC–TV) has drastically revised its content since it began in October 1970. SDC–TV may survive and be successful financially. It may also be the most notable local origination failure of all, because it now shows (5 nights a week) double-feature western movies with all the old stereotypes of good-guy-in-white-hat who hates women and Indians but loves his horse and bearded companion. And SDC–TV is operated by a black man.

Case Study Comparison

Although there were other minor problems involved in the above examples, the success or failure of local origination in each case can be traced to finances:

Canada	Viable, profitable cable system subsidizes local origination = *Success.*
Dale City	Viable, but low profit cable system could not subsidize local origination = *Failure*.
Lakewood	Non-viable, unprofitable cable system was unable to subsidize local origination = *Failure*.

San Diego Viable, profitable cable system does not subsidize independent local origination = *Anticipated failure*, but, unlike Dale City, local originator is getting advertiser support = *May succeed.*

This comparison of the four case studies is somewhat oversimplified, but the importance of operating a local origination station or channel as a subsidized or non-subsidized *business* cannot be stressed too strongly.

In summary, the success or failure of origination will depend on its income. Whether it is advertising, pay cable, foundation or CATV system support, there must be income to pay for the equipment facilities and personnel. And, in that sense, cablecasting is just like everything else.

□

CHAPTER 4

municipal regulations of cable communications

by james hudson

introduction

Local regulation of cable communications must be understood by groups working in the minority communities if such groups are to be competitive with any degree of success. In many instances, by the time community groups have acquired sufficient information as a basis for taking action, local ordinances are passed or are so near passage, that it is impossible to manage full participation. *It is therefore imperative that interested citizens move immediately to find out what is happening—both publicly and behind the scenes—on local regulations.*

Who Regulates Cable TV?

Presently, cable communications are regulated at all three levels of government—federal, state and local. But of the three, local regulations are the most comprehensive and actually determine the kind of service received by the viewer. As discussed elsewhere, the Federal Communications Commission (FCC) has exercised rather limited jurisdiction. Similarly, at the state level, only a few public utility commissions have exercised any jurisdiction.[1] All of the regulations concerning size of operation, quality, type of service and fees chargeable are determined on the local level.

How is Cable TV Regulated?

Municipal regulation is based on the city's right to regulate the use of the streets and public ways. This is because the cable system requires a complex network of coaxial cable using the public ways. Sound and pictures are transmitted over the wires instead of through the air. The most common form of municipal grant or authority to use the public ways is acquired by franchise.[2] Through franchise, the city gives the cable company the right to lay cable in the public ways. Each cable usually has a 12 to 15 channel capacity, and often cable companies will acquire permission to lay two such cables side by side. This means that a cable company operates a cable system which can broadcast on 24 to 30 TV channels. The franchise agreement sets forth the geographical area of the company's operation and the terms and conditions of such an operation.

How Are Cable Franchises Awarded?

The municipal regulatory process usually begins with a request for a franchise submitted to the municipal governing body by persons interested in cable communications. Or the city may, on its own initiative, request proposals. When the proposals are submitted, hearings are held in which interested persons submit their views on the proposed cable system.

The hearing stage is most critical in setting forth the framework of operation for the cable system. The systems proposed might be: (1) public ownership, (2) community group nonprofit ownership, (3) community group profit ownership, or (4) profit ownership. The factors which will guide any municipal action on the request are produced at the hearing. The hearing authority, in accordance with its own rules, will hear all the technical questions of the cable system, as well as questions about the type of ownership. *It is at this stage, if not before, that groups operating in minority communities must be thoroughly prepared to make presentations and push positions.*

What Is Included In a Franchise Proposal?

As noted, municipal hearings should cover all phases of cable system. Therefore, all phases of the system must be covered in the proposal. The most commonly regulated matters which are considered in hearings and usually included in franchise agreements are as follows:

(1) *Area of Operations.* A determination may be initially made to limit one company's operation to one section, district or subdistrict of the city; or, the franchise agreement may allow the cable company to operate over the entire city. In a district arrangement, the city limits one company's operation to certain districts reserving other districts for competing companies.

Note: One of the significant aspects of limiting a company's operation to a smaller area is that (1) it guards against a monopoly arrangement for the entire city area and (2) makes it possible for a company with smaller financial resources to compete for a smaller district franchise. This is particu-

[1] See Robert Winters, *Municipal Regulation of CATV – Community Antenna Television – Model Ordinances*. These states are Connecticut, Rhode Island, Vermont, Nevada, and Hawaii.

[2] The franchise agreement is usually set forth in an Ordinance. The adoption of an Ordinance usually is pursuant to municipal council rules as to notice and opportunity to be heard.

larly important for community groups who may have limited resources and might not be interested in ownership of a city-wide system.

(2) *Exclusive or Nonexclusive Agreements.* As with the restricted service area above, the nonexclusive franchise agreement seeks to prohibit a monopoly situation. Under the nonexclusive provision of the agreement, the city reserves the right to grant competing franchises. Franchises may be competitive in the same or differing areas but the economics of cable make competing franchises in the same physical area unrealistic.

(3) *Service Obligations.* Probably the most important regulatory provision from the community group's perspective is the service obligation provision. This section of the ordinance directs to whom service must be provided and on what basis. The term "service" is a very broad one—meaning not only service by the company to the viewers, but also the use of the company's communication facilities (studios, cameras, etc.) by other public interest groups. Under some provisions, the company is obligated to lease its facilities to as many diverse groups as possible.

Note: This arrangement provides community groups who may not be financially or technically able to compete in an ownership struggle, a legally assured opportunity to some broadcast time.

A commonly used provision is "free channel set-asides" for governmental use, i.e., schools, libraries, etc., or emergency uses, such as fire or police matters. A more meaningful provision is a company obligation to create a public channel.[3] This channel would be made completely available for program transmission by various public interest groups. Under such a provision, the company is also required to make broadcast facilities available and to provide technical assistance.

[3]Public Channel has been defined to mean "channels on the system which are reserved . . . for the carriage of program material provided by persons who lease the channel time and, if necessary, studio facilities, from the company for the presentation of programs . . ." Resolution, June 18, 1970, Cal. No. 414, New York City.

(4) *Percentage Fees as Municipal Levies.* The municipality has the discretionary authority to impose a fee for granting the use of city streets and public ways for the coaxial cable. The fee is usually expressed in terms of a percentage of the gross receipts. The fees may range from one percent of gross receipts to thirteen percent of gross receipts. For instance, in Illinois, the percentage fee ranges from a low 1½ percent to a high of ten percent.[4]

(5) *Pay Television.* As a concession to the pressures from the theatre industry and the television industry, many franchise agreements prohibit the transmission of "pay television" signals. However, there are non-theatrical applications of the "pay-TV" idea, such as burglar surveillance, which are now usually grouped under the term "additional services" of cable communications. "Pay television" may provide a lucrative source of channel leasing opportunities to CATV owners.

(6) *Preferential or Discriminatory Practices.* Ordinances usually provide that "The company shall not, as to rates, charges, service facilities, rules, regulations, or in any other respect make or grant any preference of advantage to any person."[5]

Note: As provisions for service, this clause is probably adequate. However, a specific provision prohibiting racial, religious or ethnic discrimination in the selection and tenure of personnel would be helpful; and also where possible the inclusion of the responsibility for affirmative action in selecting employees from among minority group persons.[6]

(7) *Television Sales and Service.* The franchise agreements uniformly prohibit companies who hold the franchise rights from engaging in the business of sales, service, leasing or maintaining television sets. Again, a more

[4]Winters, supra, p. 71.

[5]*Ordinance No. 19050*, Tacoma, Washington.

[6]See the Model Ordinance drafted by James L. Hudson, *Section IIII.*

expansive provision would be a limitation or prohibition in engaging in any allied services, including the manufacturing or servicing of telecommunications equipment.

Most ordinances are drafted to regulate only the broadcast functions[7] of cable operations. Restriction on allied activities, such as manufacturing, repair, servicing and non-broadcast uses would increase the opportunities for community groups to establish businesses to provide these services.

The above discussion is intended to give an overview of municipal regulation from a perspective that is more important for groups working within minority communities. Municipal regulations, however, include many provisions and aspects not discussed above. Such regulations include: (1) specifications as to quality of signals to be maintained, (2) required channel capacity, (3) installation fees and monthly rates the company may charge subscribers, (4) transferability of franchise, (5) duration of franchise and (6) obligation of company to "save the municipality harmless" in the event of liability or loss.

The following Ordinance should be reviewed in terms of the more technical provisions of municipal regulation and for the commonly used terms in cable communication regulation.

[7]Most Ordinances were adopted at the more limited stage of cable communications when the foreseeable uses were limited to the transmission of television signals. However, with the advent of the exploding technology in cable uses for alarm systems, written word transmission and response, emergency fire and police hook-ups and a host of other uses, many previously adopted ordinances (such as the "Pay-TV" ordinance) are rapidly becoming outdated.

Editor's Note: The following franchise ordinance is an example of the type minority groups might seek. However, actual ordinance provisions will be based on local laws, procedures, requirements, politics, and other pertinent conditions.

MODEL ORDINANCE GRANTING CABLE COMMUNICATIONS FRANCHISE

By James L. Hudson

An Ordinance Granting a Franchise to the ______________________ ___________, its Successors and Assigns, to Operate and Maintain a Cable Communications System in the City; Setting Forth Conditions Accompanying the Grant of Franchise; Providing for City Regulation and Use; and Prescribing Penalties for Violation of the Franchise Provisions.

BE IT ORDAINED BY THE COUNCIL OF THE CITY OF __________:

SECTION 100. *Short Title*. This Ordinance shall be known and may be cited as the "____________ Cable Communications Franchise Ordinance."

SECTION 101. *Definitions*. For the purposes of this Ordinance, the following terms, phrases, words, and their derivations shall have the meaning given herein.

(1) "City" is the City of ______________________________.

(2) "Council" is the City Council of ______________________.

(3) "Cable Communications," herinafter referred to as "cable system," means a system of coaxial cables or other electrical conductors and equipment used or to be used to receive television or radio signals directly or indirectly off-the-air and transmit them to subscribers and any additional service provided by the coaxial cable, including, by way of example but not limited to, burglar alarm data or other electronic intelligence transmission, facsimile reproduction, reading and home shopping.

(4) "Person" is any person, firm, partnership, association, corporation, company or organization of any kind.

(5) "Grantee" is ____________ (Name of Grantee) or anyone who succeeds ____________ (Name of Grantee) in accordance with the provisions of this Franchise.

(6) "Public Channel" is the channel on the cable system which is reserved by this Ordinance, for the carriage of program material provided by persons who lease channel time and, if necessary, studio facilities, from the Company for the presentation of programs in accordance with Section ______ of this Ordinance.

(7) "Minority Groups" means persons of African descent, Mexican-Americans, Puerto Ricans, the Oriental races and American Indians.

SECTION 102. *Grant of Nonexclusive Authority.*

(a) There is hereby granted by the City to the grantee the right and privilege to construct, erect, operate and maintain, in, upon, along, across, above, over and under the streets, alleys, public ways and public places now laid out or dedicated, and all extensions thereof, and additions thereto, in the City, poles, wires, cables, underground conduits, manholes, and other television conductors and fixtures necessary for the maintenance and operation in the City of a cable system.

(b) The right to use and occupy said streets, alleys, public ways and places for the purposes herein set forth shall not be exclusive, and the City reserves the right to grant a similar use of said streets, alleys, public ways and places, to any person at any time during the period of this Franchise.

SECTION 103. *Compliance with Applicable Laws and Ordinances.* The grantee shall, at all times during the life of this Franchise, be subject to all lawful exercise of the police power by the City and to such reasonable regulation as the City shall hereafter provide.

SECTION 104. *Territorial Area Involved.* This Franchise relates to[8]

---

[8]The area or areas of the city should be set out here.

SECTION 105. *Liability and Indemnification.*

(a) The grantee shall pay and by its acceptance of this Franchise the grantee specifically agrees that it will pay all damages and penalties which the City may legally be required to pay as a result of granting this Franchise. These damages or penalties shall include, but shall not be limited to, damages arising out of the copyright infringements and all other damages arising out of the installation, operation, or maintenance of the cable system authorized herein, whether or not any act or omission complained of is authorized, allowed, or prohibited by this Franchise.

(b) The grantee shall pay and by its acceptance of this Franchise specifically agrees that it will pay all expenses incurred by the City in defending itself with regard to all damages and penalties mentioned in subsection (a) above. These expenses shall include all out-of-pocket expenses, such as attorney fees, and shall also include the reasonable value of any services rendered by the City Attorney or his assistants or any employees of the City.

(c) The grantee shall maintain, and by its acceptance of this Franchise specifically agrees that it will maintain throughout the terms of this Franchise liability insurance insuring the City and the grantee with regard to all damages mentioned in sub-paragraph (a) above in the minimum amounts of:

(1) $_______________ for bodily injury or death to any one person, within the limit, however, of $_______________ for bodily injury or death resulting from any one accident.

(2) $_______________ for property damage resulting from any one accident.

(3) $_______________ for all other types of liability.

(d) The grantee shall maintain, and by its acceptance of this Franchise specifically agrees that it will maintain throughout the term of this Franchise a faithful performance bond running to the City, with at least two good and sufficient sureties approved by the City, in the penal sum of $_______________ conditioned that the grantee shall well and truly observe, fulfill, and perform each term and condition of this Franchise and that in case of any breach of condition of the bond, the amount thereof shall be recoverable from the principal and sureties thereof by the City for all damages proximately resulting from the failure of the grantee to well and faithfully observe and perform any provision of this Franchise.

(e) The insurance policy and bond obtained by the grantee in compliance with this section must be approved by the City Council and such insurance policy and bond, along with written evidence of payment of required premiums, shall be filed and maintained with the _______________ (City Clerk or other appropriate officer or employee of the City) during the term of this Franchise.

SECTION 106. *Prohibition of Pay TV.* The grantee is specifically barred from delivering television signals, directly or indirectly, from any pay-television source.

SECTION 107. *Color TV.* The facilities used by the grantee shall be capable of distributing color TV signals, and when the signals the grantee distributes are received in color they shall be distributed in color where technically feasible.

SECTION 108. *Signal Quality Requirements.* The grantee shall:

(a) Produce a picture, whether in black and white or in color, that is undistorted, free from ghost images, and accompanied with proper sound on typical standard production TV sets in good repair, and as good as the state of the art allows:

(b) Transmit signals of adequate strength to produce good pictures with good sound at all outlets without causing cross-modulation in the cables or interfering with other electrical or electronic systems;

(c) Limit failures to a minimum by locating and correcting malfunctions promptly, but in no event longer than ____________ hours after notice;

(d) Demonstrate by instruments and otherwise to subscribers that a signal of adequate strength and quality is being delivered.

(e) The grantee shall have the capacity to transmit over the cable system the signals of at least twelve (12) channels.

SECTION 109. *Operation and Maintenance of System.*

(a) The grantee shall render efficient service, make repairs promptly, and interrupt service only for good cause and for the shortest time possible. Such interruptions insofar as possible shall be preceded by notice and shall occur during periods of minimum use of the system.

(b) The grantee shall maintain an office in the City, which shall be open during all usual business hours, have a listed telephone, and be so operated that complaints and requests for repairs or adjustments may be received at any time.

SECTION 110. *Service Obligations.*

(a) The grantee shall receive and distribute television and radio signals which are disseminated to the general public without charge by broadcasting stations licensed by the Federal Communications Commission. All FCC regulations shall be complied with regarding the carriage of the programming of any existing or future television broadcasting station which covers the City of __________________ in its principal broadcasting area.

(b) In the operation of the cable system the grantee shall provide, on a non-discriminatory basis, a reasonable amount of free time to legally qualified candidates for public office.

(c) The grantee shall establish a public channel for programming and shall lease time and if necessary, adequate studio facilities to members of the public at rates filed and approved by the Council. Appropriate technical assistance shall also be furnished by the grantee.

SECTION 111. *Program Alteration.* All programs of broadcasting stations carried by the grantee shall be carried in their entirety as received, with announcements and advertisements and without additions.

SECTION 112. *Service to Schools.* The grantee shall provide service to public school locations and teaching stations within the City for educational purposes upon request by the City and at no cost to it or to the public school system. The grantee may at its election provide similar services without cost to private schools, including parochial or other religious schools.

SECTION 113. *Emergency Use of Facilities.* In the case of any emergency or disaster, the grantee shall, upon request of the City Council, make available its facilities to the City for emergency use during the emergency or disaster period.

SECTION 114. *Other Business Activities.*

(a) Neither the grantee hereunder nor any shareholder of the grantee shall engage in the business of selling, repairing, or installing television receivers, radio receivers, or accessories for such receivers within the City of ________ ________ during the term of this Franchise and the grantee shall not allow any of its shareholders to so engage in any such business.

(b) The grantee shall not enter the additional services of burglar alarm data or other electronic intelligence transmission, facsimile reproduction, reading and home shopping until the grantee has requested authority to provide such services from the Council.

SECTION 115. *Safety Requirements.*

(a) The grantee shall at all times employ ordinary care and shall install and maintain in use commonly accepted methods and devices for preventing failures and accidents which are likely to cause damage, injuries, or nuisances to the public.

(b) The grantee shall install and maintain its wires, cables, fixtures, and other equipment in accordance with the requirements of the City electric code and in such manner that they will not interfere with any installations of the City or of a public utility serving the City.

(c) All structures and all lines, equipment, and connections in, over, under, and upon the streets, sidewalks, alleys, and public ways or places of the City, wherever situated or located, shall at all times be kept and maintained in a safe, suitable, substantial condition, and in good order and repair.

(d) The grantee shall maintain a force of one or more resident agents or employees at all times and shall have sufficient employees to provide safe, adequate, and prompt service for its facilities.

SECTION 116. *New Developments.* It shall be the policy of the City liberally to amend this Franchise, upon application of the grantee, when necessary to enable the grantee to take advantage of any developments in the fields of transmission of television and radio signals which will afford it an opportunity more effectively, efficiently, or economically to serve its customers. Provided, however, that this Section shall not be construed to require the City to make any amendment or to prohibit it from unilaterally changing its policy stated herein.

SECTION 117. *Conditions on Street Occupancy.*

(a) All transmissions and distribution structures, lines, and equipment erected by the grantee within the City shall be so located as to cause minimum interference with the proper use of streets, alleys, and other public ways and places, and to cause minimum interference with the rights and reasonable convenience of property owners who join any of the said streets, alleys or other public ways and places.

(b) In case of disturbance of any street, sidewalk, alley, public way, or paved area, the grantee shall, at its own cost and expense and in a manner approved by the ______________________ (Director of Public Works or other appropriate official), replace and restore such street, sidewalk, alley, public way, or paved area in as good a condition as before the work involving such disturbance was done.

(c) If at any time during the period of this Franchise the City shall lawfully elect to alter or change the grade of any street, sidewalk, alley, or other public way, the grantee, upon reasonable notice by the City, shall remove, relay, and relocate its poles, wires, cables, underground conduits, manholes, and other fixtures at its own expense.

(d) Any poles or other fixture placed in any public way by the licensee shall be placed in such manner as not to interfere with the usual travel on such public way.

(e) The grantee shall, on the request of any person holding a building moving permit issued by the City, temporarily raise or lower its wires to permit the moving of buildings. The expense of such temporary removal or raising or lowering of wires shall be paid by the person requesting the same, and the grantee shall have the authority to require such payment in advance. The grantee shall be given not less than forty-eight (48) hours' advance notice to arrange for such temporary wire changes.

(f) The grantee shall have the authority to trim trees upon and overhanging streets, alleys, sidewalks, and public ways and places of the City so as to prevent the branches of such trees from coming in contact with the wires and cables of the grantee, except that at the option of the City, such trimming may be done by it or under its supervision and direction at the expense of the grantee.

(g) In all sections of the City where the cables, wires, or other like facilities of public utilities are placed underground, the grantee shall place its cables, wires or other like facilities underground to the maximum extent that existing technology reasonably permits the grantee to do so.

SECTION 118. *Preferential or Discriminatory Practices Prohibited.* The grantee shall not, as to rates, charges, service, service facilities, rules, regulations, or in any other respect, make or grant any undue preference or advantage to any person, nor subject any person to prejudice or disadvantage. The grantee shall not discriminate against any person on racial, religious, or ethnic basis in the selection and tenure of its employees. The grantee shall also take affirmative actions to recruit employees from members of minority groups.

SECTION 119. *Removal of Facilities Upon Request.* Upon termination of service to any subscriber, the grantee shall promptly remove all its facilities and equipment from the premises of such subscriber upon his request.

SECTION 120. *Transfer of Franchise.* The grantee shall not transfer this Franchise to another person without prior approval of the City by ordinance.

SECTION 121. *Transactions Affecting Ownership of Facilities.*

(a) In order that the City may exercise its option to take over the facilities and property of the cable system authorized herein upon expiration or forfeiture of the rights and privileges of the grantee under this Franchise, as is provided for herein, the grantee shall not make, execute, or enter into any deed, deed of trust, mortgage, conditional sales contract, or any loan, lease, pledge, sale, gift or similar agreement concerning any of the facilities and property, real or personal, of the cable business without prior approval of the City Council upon its determination that the transaction proposed by the grantee will not be inimical to the rights of the City under this Franchise. Provided, however, that this section shall not apply to the disposition of worn out or obsolete facilities or personal property in the normal course of carrying on the cable business.

(b) Except as provided for in subsection (a) above, the grantee shall at all times be the full and complete owner of all facilities and property, real and personal, of the cable business.

SECTION 122. *Change of Control of Grantee.* Prior approval of the City Council shall be required where ownership or control of more than 30% of the right of control of grantee is acquired by a person or group of persons acting in concert, none of whom already own or control 30% or more of such right of control, singularly or collectively. By its acceptance of this Franchise the grantee specifically grants and agrees that any such acquisition occurring without prior approval of the City Council shall constitute a violation of this Franchise by the grantee.

SECTION 123. *Filings and Communications with Regulatory Agencies.* Copies of all petitions, applications and communications submitted by the grantee to the Federal Communications Commission, Securities and Exchange Commission, or any other federal or state regulatory commission or agency having jurisdiction in respect to any matters affecting cable operations authorized pursuant to this Franchise, shall also be submitted simultaneously to the City Council.

SECTION 124. *City Rights in Franchise.*

(a) The right is hereby reserved to the City or the City Council to adopt, in addition to the provisions contained herein and in existing applicable ordinances, such additional regulations as it shall find necessary in the exercise of the police power; provided that such regulations, by ordinance or otherwise, shall be reasonable and not in conflict with the rights herein granted.

(b) The City shall have the right to inspect the books, records, maps, plans, income tax returns, and other like materials of the grantee at any time during normal business hours.

(c) The City shall have the right, during the life of this Franchise, to install and maintain free of charge upon the poles of the grantee any wire and pole fixtures necessary for a police alarm system, on the condition that such wire and pole fixtures do not interfere with the cable operations of the grantee.

(d) The City shall have the right to supervise all construction or installation work performed subject to the provisions of this Franchise and make such inspections as it shall find necessary to insure compliance with the terms of this Franchise and other pertinent provisions of law.

(e) At the expiration of the term for which this Franchise is granted, or upon its termination and cancellation, as provided for herein, the City shall have the right to require the grantee to remove at its own expense all portions of the cable system from all public ways within the City.

(f) At the expiration of the term for which this Franchise is granted, or upon its termination and cancellation, as provided for herein, the City, at its election, and upon the payment of ________________ [9] to the grantee, shall have the right to purchase and take over the cable system in its entirety. The above price shall not include, and the grantee shall not receive anything for the valuation of any right or privilege appertaining to it under this Franchise.

---

[9] Provide here for method of evaluation such as original cost less accumulated depreciation, fair market value, replacement value, arbitration, etc., where charter or statutes are not controlling.

Upon the exercise of this option by the City and its service of an official notice of such action upon the grantee, the grantee shall immediately transfer to the City possession and title to all facilities and property, real and personal, of the cable business, free from any and all liens and encumbrances not agreed to be assumed by the City in lieu of some portion of the purchase price set forth above; and the grantee shall execute such warranty deeds or other instruments of conveyance to the City as shall be necessary for this purpose. The grantee shall make it a condition of each contract entered into by it with reference to its operations under this Franchise that the contract shall be subject to the exercise of this option by the City and that the City shall have the right to succeed to all privileges and obligations thereof upon the exercise of such option. Provided, however, that the City shall have the right unilaterally to increase the purchase price provided for above, should it so elect, by an ordinance amendatory hereto. But such right shall not be construed as giving the grantee a right to any price in excess of that set forth above.

(g) After the expiration of the term for which this Franchise is granted, or after its termination and cancellation, as provided for herein, the City shall have the right to determine whether the grantee shall continue to operate and maintain the cable system pending the decision of the City as to the future maintenance and operation of such system.

SECTION 125. *Maps, Plats, and Reports.*

(a) The grantee shall file with the City Clerk true and accurate maps or plats of all existing and proposed installations.

(b) The grantee shall file annually with the City Clerk not later than sixty (60) days after the end of the grantee's fiscal year, a copy of its report to its stockholders (if it prepares such a report), an income statement applicable to its operations during the preceding 12 months period, a balance sheet, and a statement of its properties devoted to cable operations, by categories, giving its investment in such properties on the basis of original cost, less applicable depreciation. These reports shall be prepared or approved by a certified public accountant and there shall be submitted along with them such other reasonable information as the City Council shall request with respect to the grantee's properties and expenses related to its cable operations within the City.

(c) The grantee shall keep on file with the City Clerk a current list of its shareholders and bondholders.

SECTION 126. *Payment to the City.* The grantee shall pay to the City annually the amount of ______________ or an amount equal to ______________ percent of the annual gross operating revenues taken in and received by it on all retail sales of television signals within the City during the year, whichever amount is greater for the use of the streets and other facilities of the City in the operation of the cable system and for the municipal supervision thereof. This payment shall be in addition to any other tax or payment owed to the City by the grantee.

SECTION 127. *Forfeiture of Franchise.*

(a) In addition to all other rights and powers pertaining to the City by virtue of this Franchise or otherwise, the City reserves the right to terminate and cancel this Franchise and all rights and privileges of the grantee hereunder in the event that the grantee:

(1) Violates any provision of this Franchise or any rule, order, or determination of the City or City Council made pursuant to this Franchise, except where such violation, other than of Section 122 or subsection (2) below, is without fault or through excusable neglect;

(2) Becomes insolvent, unable or unwilling to pay its debts, or is adjudged a bankrupt;

(3) Attempts to dispose of any of the facilities or property of its cable business to prevent the City from purchasing same, as provided for herein;

(4) Attempts to evade any of the provisions of this Franchise or practices any fraud or deceit upon the City; or

(5) Fails to begin (complete) construction under this Franchise before ______________________________ .

(b) Such termination and cancellation shall be by ordinance duly adopted after __________ days notice to the grantee and shall in no way affect any of the City's rights under this Franchise or any provision of law. In the event that such termination and cancellation depends upon a finding of fact, such finding of fact as made by the City Council or its representative shall be conclusive. Provided, however, that before this Franchise may be terminated and cancelled under this Section, the grantee must be provided with an opportunity to be heard before the City Council.

SECTION 128. *City's Right of Intervention.* The grantee agrees not to oppose intervention by the City in any suit or proceeding to which the grantee is a party.

SECTION 129. *Further Agreement and Waiver by Grantee.* The grantee agrees to abide by all provisions of this Franchise, and further agrees that it will not at any future time set up as against the City or the City Council the claim that the provisions of this Franchise are unreasonable, arbitrary, or void.

SECTION 130. *Duration and Acceptance of Franchise.*

(a) This Franchise and the rights, privileges, and authority hereby granted shall take effect and be in force from and after final passage hereof, as provided by law, and shall continue in force and effect for a term of __________ years, provided that within __________ days after the date of the passage of this ordinance the grantee shall file with the City Clerk its unconditional acceptance of this Franchise and promise to comply with and abide by all its provisions, terms, and conditions. Such acceptance and promise shall be in writing duly executed and sworn to, by or on behalf of the grantee before a notary public or other officer authorized by law to administer oaths.

(b) Should the grantee fail to comply with subsection (a) above it shall acquire no rights, privileges, or authority under this Franchise whatever.

SECTION 131. *Erection, Removal, and Common User of Poles.*

(a) No poles or other wire-holding structures shall be erected by the grantee without prior approval of the City Council with regard to location, height, type and any other pertinent aspect. However, the location of any pole or wire-holding structure of the grantee shall be removed or modified by the grantee at its own expense whenever the City Council determines that the public convenience would be enhanced thereby.

(b) Where poles or other wire-holding structures already existing for use in serving the City are available for use by the grantee, but it does not make arrangements for such use, the City Council may require the grantee to use such poles and structures if it determines that the public convenience would be enhanced thereby and the terms of the use available to the grantee are just and reasonable.

(c) Where the City or a public utility serving the City desires to make use of the poles or other wire-holding structures of the grantee but agreement therefor with the grantee cannot be reached, the City Council may require the grantee to permit such use for such consideration and upon such terms as the Council shall determine to be just and reasonable, if the Council determines that the use would enhance the public convenience and would not unduly interfere with the grantee's operations.

SECTION 132. *Rates.*

(a) The rates and charges for television and radio signals distributed hereunder shall be fair and reasonable and no higher than necessary to meet all costs of service (assuming efficient and economical management), including a fair return on the original cost, less depreciation, of the properties devoted to such service (without regard to any subsequent sale or transfer price or cost of such properties).

(b) The City Council shall have the power, authority, and right to cause the grantee's rates and charges to conform to the provisions of subsection (a) hereof, and for this purpose, it may deny increases or order reductions in such rates and charges when it determines that in the absence of such action on its part, the grantee's rates and charges or proposed increased rates and charges will not conform to the said subsection (a).

(c) By its acceptance of this Franchise the grantee specifically grants and agrees that its rates and charges to its subscribers for television and radio signals shall be fair and reasonable and no higher than necessary to meet all its necessary costs of service (assuming efficient and economical management), including a fair return on the original cost, less depreciation, of its properties devoted to such service (without regard to any subsequent sale or transfer price or cost of such properties).

(d) By its acceptance of this Franchise the grantee further specifically grants and agrees that the City Council shall have the power, authority, and right to cause the grantee's rates and charges to conform to the provisions of subsection (c) hereof, and for this purpose, the Council may deny increases or order reductions in such rates and charges when it determines that in the absence of such action on its part, the grantee's rates and charges or proposed increased rates and charges will not conform to the said subsection (c).

(e) However, no action shall be taken by the City Council with respect to the grantee's rates under this Section until the grantee has been given reasonable notice thereof and an opportunity to be heard by the Council with regard thereto.

(f) The following rates and charges are hereby authorized for service under this Franchise and shall not be changed by the grantee without prior approval by the City Council:

(1) Initial tap-in and connection charges: $8.00;

(2) Monthly rates: $5.00 per month.

(g) The grantee shall receive no deposit, advance payment, or penalty from any subscriber or potential subscriber without approval of the Council.

(h) The grantee shall receive no consideration whatsoever for or in connection with its service to its subscribers other than in accordance with this Section.

(i) If in the future, the State of ________________ regulates the rates of the grantee for the service provided for in this Franchise, this Section shall be of no effect during such state regulation to the extent of any conflict therewith.

SECTION 133. *Flow-through of Refunds.*

(a) If during the term of this Franchise the grantee receives refunds of any payments made for television or radio signals, it shall without delay notify the City Council, suggest a plan for flow through of the refunds to its subscribers, and retain such refunds pending order of the Council. After considering the plan submitted by the grantee, the Council shall order the flow through of the refunds to the grantee's subscribers in a fair and equitable manner.

(b) By its acceptance of this Franchise the grantee specifically grants and agrees that if, during the term hereof, it receives refunds of any payments made for television or radio signals, it shall without delay notify the City Council, suggest a plan for flow through of the refunds to its subscribers, retain the refunds pending order of the Council, and flow through such refunds in accordance with the order of the Council.

SECTION 134. *Subscriber Refunds on Termination of Service*. If any subscriber of the grantee of less than three (3) years terminates service because of the grantee's failure to render service to such subscriber of a type and quality provided for herein, or if service to a subscriber of less than three (3) years is terminated without good cause or because the grantee ceases to operate the cable business authorized herein for any reason, except expiration of this Franchise, the grantee shall refund to such subscriber an amount equal to the initial tap-in and connection charges paid by him divided by 36 and multiplied by a number equal to 36 minus the number of months the subscriber has been on the system.

SECTION 135. *Publication Costs*. The grantee shall assume the cost of publication of this Franchise as such publication is required by law and such is payable upon the grantee's filing of acceptance of this Franchise.

SECTION 136. *Separability*. If any section, subsection, sentence, clause, phrase, or portion of this ordinance is for any reason held invalid or unconstitutional by any court of competent jurisdiction, such portion shall be deemed a separate, distinct, and independent provision and such holding shall not affect the validity of the remaining portions hereof.

SECTION 137. *Ordinances Repealed*. All ordinances of parts of ordinances in conflict with the provisions of this ordinance are hereby repealed. These are: ______________________________

____

____

SECTION 2

workshop sponsors

National Business League
Washington, D.C.
Berkeley Burrell, President

National Council for Equal Business Opportunity
Washington, D.C.
Ben Goldstein, Director

Cooperative Assistance Fund
Washington, D.C.
Edward Sylvester, Director

Office of Minority Business Enterprise
U.S. Department of Commerce
Washington, D.C.
Jay Leanse, Deputy Director

workshop conveners

Black Efforts for Soul In Television
William Wright

Urban Communications Group
Ted Ledbetter

The Urban Institute
Francille Rusan
Charles Tate

WORKSHOP REPORT

minority business opportunities in cable television

introduction

At a four-day workshop called "What You See Is What *To* Get," community development leaders and practioners, entrepreneurs, and economic planners discovered that the mass media, long closed to minority ownership and control, has been forced open by cable television's pending expansion into the nation's big cities.

Convened by The Urban Institute, the Urban Communications Group and Black Efforts for Soul in Television (BEST) the workshop met June 24-27, 1971 at the Sonesta Hotel in Washington, D.C. Most of the 110 participants were representatives of Community Development Corporations (CDC's) and other community based organizations. Others attending were communications experts, foundation, personnel, representatives from national organizations concerned about minorities and communications, and staff personnel from federal agencies involved in communications policy and minority economic development.

Participants came from Boston, Buffalo, Brooklyn, Baltimore, Chicago, Cleveland, Dayton, Detroit, Durham, Kansas City, Newark, Harlem, Philadelphia, Los Angeles, Roanoke, Rochester, Milwaukee, Memphis and Washington, D.C.

Charles Tate of The Urban Institute indicated that the workshop's primary objective was "to make minority economic developers, bankers, and technical assistance organizations collectively aware of the economic potential of cable communications technology, and thus stimulate minority leaders to develop cable projects in their own communities."

The key issue, then, is whether black people will be able to own part of the cable industry or left in the situation of demanding jobs and access to a system totally owned by somebody else.

The workshop featured panel discussions by persons experienced in the legal and business aspects of cable television. A tone of urgency was set by the main speakers who stressed the need for immediate involvement in this developing industry and the importance of understanding the political strategies needed to get minority control.

The opening speaker, William Wright, president of Black Efforts for Soul in Television (BEST), emphasized that efforts to challenge the licenses of broadcast stations on the issue of discrimination have fallen far short of the goal.

After battling the networks in Columbus, Ohio, "the visual complexions of TV began to change, but it still didn't change in terms of ownership and control of programming," he said.

"We've missed the boat in owning broadcast stations." Wright concluded, "Cable television is the last stronghold as far as getting in on the communications medium."

Not only is cable "the last stronghold," but it may also be the best opportunity, according to the facts presented by workshop panelists.

Workshop participants—aware of CATV's business possibilities as well as the potential for a strong local communications system—stressed that these opportunities must be realized now before white businesses take complete control of cable television's development in the cities.

The Federal Communications Commission, under pressure from the powerful broadcast lobby, has banned the importation of distant signals by cable companies in the Top 100 markets. This ban means a cable company in Washington cannot bring in New York stations. On August 6, 1971, the FCC sent Congress a long explanation of its proposed plan to lift the ban. It is expected that the decision to open these top television markets will be made before the end of 1971. In the meantime, white companies are scrambling for the franchise rights to build systems as soon as the ban is lifted.

Citing examples of the growing interest in cable television among the major white corporations, Wright urged immediate action. He said:

> When Howard Hughes gets involved in anything, you know it's money. And when Time-Life starts selling off their television interests to begin looking into cable companies in this country and people who are connected with them, then you know it's important. . .
>
> If you do any foot dragging when you go back to your respective economic development groups, you'll be out of the ball game.
>
> Once those franchises are let, you can almost forget it. Put someone on it full time, on a day-to-day basis; otherwise it will be too late.
>
> If you don't get involved in this arena, the only thing you'll be able to do on cable is sing and dance. . .and that will be prostitution again.

The four-day workshop began with a tour of an operating cable facility in Reston, Virginia. During the following three days, panel presentations, workshop discussion groups, luncheon and dinner speakers and film and video-tape showings were conducted. Roy Lewis, assistant director of Notre Dame University's Project Reach, presented two films, one on the nation of Islam and the other on poet Gwendolyn Brooks. Another black film-maker, Vernard Gray, director of

Fides House Communications Workshop in Washington, D.C., showed video-tapes made on the streets of D.C. using ½" Sony Portapak equipment which is versatile, inexpensive video equipment ideal for use by community groups. These video tapes, covering a variety of information, education, and entertainment programs could be shown over a community-wide, community controlled cable system.

Owners and operators of cable systems, and specialists with a working knowledge of cable technology and business management contributed heavily to the panel presentations and working sessions. Organizations involved in providing financial aid and technical assistance to minority groups and entrepreneurs were active participants in all of the sessions. These were The National Council for Equal Business Opportunity, The National Business League, National Progress Association for Economic Development, Opportunity Funding Corporation, Cooperative Assistance Fund, Community Relations Service, National Banking Association, National Mortgage Bankers Association, Community Film Workshop Council, Cummins Engine Foundation, Urban Communications Group, Black Efforts for Soul in Television, Fides House Communications Center, and the Office of Minority Business Enterprise, U.S. Commerce Department.

The following report is not a verbatim account of all Workshop activities. Major presentations and other selected materials are presented here in summary form to provide the general audience with both the information and the sense of priorities expressed in the Workshop.

Editor

CATV background and technology

PANELISTS:

GARY CHRISTENSEN, General Counsel, National Cable Television Association; Washington, D.C.

STAN GERENDASY, Consulting Engineer, Ford Foundation; New York, N.Y.

TED LEDBETTER, President, Urban Communications Group; Washington, D.C.

SOL SCHILDHAUSE, Chief CATV Bureau, Federal Communications Commission; Washington, D.C.

Workshop participants generally agreed that cable television will dominate the communications industry within the next ten years, if it's growth is not impeded by broadcast television, movie, telephone and newspaper industries or the FCC. It is projected to be a 4.4 billion dollar industry providing almost two million new jobs by 1980.

Gary Christensen, the first panelist to speak during the discussion of cable background and technology, said that "the prospects for growth depend on the regulations made by the FCC." Christensen, General Counsel for the National Cable Television Association (NCTA) told the audience that the NCTA "has become a political and legal organization because of the necessity to deal with these issues. I would suggest that you resist any attempt to put a freeze on cable."

He said that interested groups or businessmen "should attempt to get franchises for urban areas or any other areas. Now is the time for you to get these franchises."

Franchises are awarded by local governments, usually the city council. The awards are usually made after public hearings to determine the contents of the ordinance. Christensen pointed out that a bidder must understand cable technology or he won't know what to demand in the ordinance governing his franchise.

One of the key provisions in a franchise ordinance is the ruling on distant signal importation. He advised, "The more distant signals you can get the better." Most existing cable systems established their sales records by being able to promise customers channels from far away cities.

A second important franchise provision involves copyright laws. Legislation now before Congress will make cable companies subject to these laws. "If it's only money, you can pay the money," he said, "but the problem is that copyright proprietors won't let cable companies buy copyrights for movies and other shows on broadcast television.

"Make sure, if you can, to get some statement of public interest by the municipality which would advocate the abolition or at least the limitation of the right of the copyright proprietor to withhold the copyright from you."

A third concern is the definition of the market area covered by the franchise. "The smaller the definition of the market area, the better off you are," he said. In a city, the cost of constructing a cable company will be far greater than the cost in rural areas, but since there are more people concentrated in a city, it is possible to have a larger number of customers within a small section of the city. Mr. Christensen, as well as later conference panelists and speakers, opposed plans to issue only one franchise for an entire city.

Christensen, questioned by Ledbetter on what position his organization took on minority ownership, said he could not promise an NCTA policy supporting community control or minority ownership, but his own belief is that "cable should be owned by the people in the community where it exists."

He added, "If you can't own it, you should go into partnership. If you can't go into partnership, you should have access to at least one channel for your own communications needs."

Issue Is Money

A dimmer view of future minority involvement in mass media was given by Sol Schildhause, CATV bureau chief for the FCC.

"Now to bring this down to earth. What does all this mean to minorities or black people?" he asked, adding that "minorities is getting to be a code word for black people."

"Over-the-air TV is clearly white television. Blacks can't get on, they can't get in, and they're always talking about and advertising things that must make the black man feel like a distant spectator."

On black ownership of cable TV he asked, "Is cable TV likely to be any better? Maybe. And that's a cautious maybe."

Noting that only 10 percent of the homes that have television sets are now hooked up to cable, he observed that the cable industry, unlike the broadcast industry is "not occupied."

"It (cable) may also end up completely white," he added because "it's a business that costs money. It's a capital intensive business, and if there's one thing blacks aren't long on, it's money."

He suggested that the main hope for black ownership was through community pressure on local governments to insure franchise grants for black corporations.

Local Origination

Stan Gerendasy, consulting engineer for the Ford Foundation, was unenthusiastic about the development of locally originated programs. He said that "cable subscribers pay for two things: additional channels that people could not get and clearer pictures."

He said the most important demand in cable is that there be a maximum number of channels because the promise of more channels will attract customers and also because channels would help insure black access to at least one channel.

Explaining the white image and program domination of broadcast television, Mr. Gerendasy said that in today's television industry, "because there are only a limited number of channels available in each city, the people who own those channels, or control them, must make the maximum use of those channels from an economic point of view and appeal to mass audiences which is why minorities have been left out. Not only minority races, not only blacks, but minority tastes, minority interests, are not serviced because of the economics of channel scarcity."

He added to his explanation of racism in broadcast television that "cable in its ability to expand this channel capacity does offer the opportunity for blacks and other interest groups to have their own channels and to get into the media."

Before opening the discussion to questions from the audience, Ted Ledbetter emphasized that the issue of a cable system's right to import distant signals is a key issue. The ability to bring in these distant signals is an incentive for subscribers to pay a monthly rate needed by the system to make any profit. This ability is also one of the main sources of contention between cable and broadcast TV. However, the FCC is beginning to place more emphasis on local program origination and less on the importation of distant signals as the main issue of consideration in the expansion of cable systems into the Top 100 markets.

The audience focused on the local origination issue during the question and answer period.

In addition to transmitting programs brought in by antenna from existing broadcast stations, cable television can put on programs of its own. This is local origination or cablecasting and it's the area that workshop conveners believe has the most potential for minority communications.

But Gerendasy opposed putting emphasis on this concept.

"There is no way to convince the guy who writes the check (for financing) that local origination will draw customers," he said in response to a claim from the audience that local origination will draw customers.

He said there were no studies available to prove that locally originated programs are financially successful—meaning that they attract subscribers who pay for cable television in order to get these local programs.

The reply from a participant in the audience was that local programming has never been tried in the black community, therefore, there was no way to tell whether it was profitable or not. The participant added that there are two independent UHF stations in New York "making it on local origination."

Schildhause pointed out that the two stations in New York were for "Spanish-speaking people," and then, to the amusement of much of the audience, added that "the black man speaks English, he's comfortable with the language."

"You can't find enough material to fill station time," he said and cable companies who want to put local shows on a channel usually use material that "someone else does and pays to have put on."

Bill Wright speaking from the floor, concluded at the end of the panel discussion that "there's got to be a model set up somewhere, whether it's the city where you live or not" to begin black involvement in cable television. The first step in getting involved is getting the franchise; and the franchising panel, one of the longest and liveliest of the workshop sessions, included panelists with firsthand knowledge of what franchising is really about.

franchising

PANELISTS:

THOMAS ATKINS, Attorney and City Councilman; Boston, Massachusetts.

JAMES HUDSON, Attorney, Hudson and Leftwich; Washington, D.C.

STEVEN RIVKIN, Attorney; Washington, D.C., former Counsel to Sloan Cable Commission.

FRANCILLE RUSAN, Urban Institute, Wash. D.C.

STUART SUCHERMAN, Communications Program Officer, Ford Foundation; New York, New York.

Francille Rusan, workshop co-ordinator and moderator for the Franchising Panel explained before opening the discussion that the presentations would describe "the things you need before you decide to go after a franchise for cable."

At many franchise hearings only one bidder shows up and neither the city government nor the general public is knowledgeable about cable television issues. A franchise is "a city ordinance or grant from a municipality to a company to use city property or public ways," attorney James Hudson explained in the opening presentation. In cable television the "major regulatory body is the city council or municipal body" which determines "the essential things concerning CATV operation: the size of the system, quality and type of service." He added that the ability of the city to make those regulations is based on "the city's right to control its public ways." A permit must be issued before a person can use the streets and sidewalks for profit.

Hudson's carefully outlined presentation on franchising indicated that the first step was attending the public hearing, usually conducted by the city council.

"The public hearing stage determines the kinds of things that are included in a franchise," he said. Among the provisions that are considered at the hearing are the market area allotted, the operating rights (exclusive or nonexclusive), leasing requirements, the city's share of the profits, and the regulation of the sale of equipment to the cable company.

"I suggest that the best way for the franchise to be granted is in districts rather than citywide," Hudson said. District franchises offer better opportunities for community groups to own their own company, because a smaller district will require a smaller financial outlay than a franchise for the entire city. An ordinance dividing the city into smaller districts also "guards against a monopoly—one company with a large amount of money moving in and taking over the entire operation."

The second ordinance issue Hudson discussed was whether the franchise is "exclusive or nonexclusive." In a nonexclusive franchise, "others can come in and can bid on a franchise even if it has been awarded," he said. That would mean another bidder could get the operating rights to the same area for which the franchise had been awarded. The third issue outlined was the lease provisions. Hudson said the ordinance provisions should require the cable company to make time available at a reasonable rate to community interest groups or to set aside a total channel solely for the use of public groups for a reasonable charge.

The fourth issue was "how much the city gets out of it." He told the audience that "the city's big stake is the fee. The city can collect a percentage of the company's profits on an increasing scale under most franchise agreements. If a community group doesn't own the franchise, it can request feedback of the fee for community development."

He added that the ordinance should also include nondiscriminatory employment provisions.

A sixth issue was the regulation of sales and services of other kinds of equipment by the company. He suggested that the ordinance include a provision that would prohibit the company from selling equipment such as television sets and from providing TV repair service.

Hudson observed that "most regulations deal only with the broadcast issues and not with current technology. There are other uses of cable TV besides broadcasting, including burglar and fire alarms, and there should be a prohibition of cable systems owning these allied services."

Hudson concluded that "the hearing stage is probably the most critical stage of the development. During this phase, community groups should let people know you know what's going on." "This is the time," he said, "to apply political pressure."

Behind Closed Doors

The three other panelists pointed out, however, that most political decisions are made behind closed doors and that by the time hearings are held on any issue the decision has already been made.

Stuart Sucherman, Ford Foundation communications program officer, offered this advice: "I think you've got to get in on the process before the public hearings. . .making yourself known and telling people you know what's going on."

He pointed out that most hearings are far from public and usually go on between individuals.

"Make people know you are there and that it's not going to be decided behind closed doors. Public hearings are usually held after a decision has been made."

Sucherman suggested that community groups first "find out something about your local government; find out who in that city is going to make that franchise decision." He concluded, "It is important to understand the political process and get a handle on that before you do anything else."

Boston City Councilman Tom Atkins, one of the most knowledgeable elected officials in the country about cable television, went even farther in his presentation. He told the group that "you're not likely to hear much about" cable television at a city council meeting and "the people who have granted franchises know very little about cable television."

He suggested that no more franchises should be awarded until the local political climate is changed and ordinance provisions enacted that would make it more likely for minorities to participate on a fair basis.

Atkins predicted that cable television will become a nationwide "communications highway" and that present franchises are awarded for too much of that highway, therefore, the market areas should be decreased. Because cable will include many nonvideo functions, he said that there should be one franchise for the video operations and a separate franchise for the nonvideo, an area usually overlooked. The nonvideo functions include the use of computers that can be linked to the "cable highway" and will allow the cable subscriber to use his television for personal research. He noted that the nonvideo application of cable "can potentially have a tremendous development impact on a community."

Atkins suggested that franchises be limited to three years, adding that financing should not be a problem because of the huge profit potential of cable.

In a statement that proved to be quite controversial, Atkins said, "I guarantee that you could take a monkey and give him a cable television franchise for a part of New York City, for Boston, for Detroit, for Chicago, and direct him toward any bank in those cities, and they'd give him the money. That's how profitable this media is going to be. He'd have no trouble convincing the most skeptical loan officer. No trouble."

"If you've got the money, you can buy the knowledge. And if you've got the franchise, you can get the money."

Atkins concluded his remarks with a series of questions about the issue of community control. He told the audience that in the next ten years there will be a drive for community control.

"But what is the community going to be asked or allowed to control. Will the community control the franchise? All of it? Will the community control the video aspect of a franchise only? Or will the community control a channel? Or, more likely, will the community control a part of a piece of a channel?

"Community control has to be understood, and has to be defined specifically and precisely before the fact. And that's something else you won't hear much about in those city council deliberations."

Steve Rivkin, a private attorney and former cable communications counsel to the Sloan Commission, followed Atkins with another view on the political problems involved in getting cable franchises. Rivkin, who is now communications counsel for Youth Organizations United (YOU) placed the responsibility on the federal government to insure black acquisition of franchises.

If the FCC lifts the ban on importing distant signals into big cities, cable will probably flourish in the Top 100 markets. The FCC has also promised to issue guidelines for federal-state-local relations at the time the ban is lifted, Rivkin pointed out. But he contended that there is nothing in the guidelines, as drafted at present, that "helps minority groups get the footing in the cable field that they deserve." Rivkin said that "the promised requirements of adequate channel capacity or of nondiscriminatory access—all these insure is the possibility of a casual, unsubsidized, ill-defined opportunity for some community gab-fests which I doubt will be adequate to service the needs felt by everyone here today."

He suggested that pressure be put on Congress for federal guidelines in the issuance of local franchises that insure minority opportunity for cable ownership. Rivkin concluded that "all of you should begin today to focus your concerns for cable, not only on city hall, but here in Washington, where the true character of the industry is about to be cast in concrete."

During the question and answer period, Ed Lloyd, who owns a cable franchise in East Orange, N.J., disputed Atkins on the ease of obtaining bank financing for cable system construction.

Lloyd declared "If he (Atkins) can show me any bank or anyone who would finance a CATV system for three years and you only have a three-year franchise—I'd like to see that bank or that group that would finance me in that.

"It takes seven years to pay out your debt. No one is going to give you money for a franchise predicated on three years." He added that it takes more than three years for a cable company to break even and "for you all to think that we can survive under a three-year franchise—well, it's ridiculous. . ."

After an exchange with Atkins on whether he had followed good business procedures in his attempts to secure a loan, Lloyd told the audience that "going into cable television, you have to have a cash flow. This is what the banks or any investment company look at. . .when you finance a CATV system, no one's going to give you a dime unless they can see where their money is going to come back and how you're going to pay it back." He said he did a study of the Roxbury and Dorchester areas in Boston and it would take about $10 million to build a cable system in those areas.

Lloyd's East Orange franchise is for 25 years, but it was suggested by the panel moderator, Ted Ledbetter, that "as long as black people are involved, it's going to be difficult to get that money, no matter how long that franchise term is."

"it's time we used our heads"

Tom Atkins

When Tom Atkins, the 35-year-old black city councilman from Boston rose to speak on the second night of the workshop, the after-dinner conversations stilled as everyone came back to the main issue—how can black people win control of community cable television.

"We've been running around in circles for years trying to meet the rules of a game when the rules were set up in part to exclude us," Atkins pointed out. "It's time we used our heads."

"Discussing the complex questions cable television presents to local communities, questions of who will actually manage the system, who will build it and how can the community have an impact on programming, Atkins offered a solution to the problem: creating a joint venture between the city, the community served, and a private corporation in partnership.

Atkins described this three-way limited partnership plan as insurance that the community will have a significant say in the management of the system whether the corporate partner is white or black.

"I would hope that the private corporation would itself be drawn from the community. I would hope that it would be an indigenously-owned community corporation," Atkins said, "but I don't equate indigenously-owned with community interests, because they're not always the same. There are a lot of bad black businessmen.

"Even the bad black businessman ought to be given a chance to operate. For years we've given bad white businessmen the chance to operate. And we're going to have to watch them."

Atkins' proposed joint venture partnership would work in the following way. The city would be a limited partner owning 15 to 20 percent. The community nonprofit corporation would own 15 to 20 percent; and the profit-making corporation, the general partner, would have 60 to 70 percent.

Atkins explained, "In my scheme the city must first divide or subdivide itself into franchise areas with none to exceed a stated percentage, 10 to 15 percent. You wind up with a number of franchise possibilities within the city, whether the percentage is 10 percent in which case we'd have at least 10, or whether it's 15 percent.

"The city would also have to set up the mechanism for selecting both the community participation structure and the private profit-making participation structure," he suggested. The city's only role, as far as community participation is concerned, would be to set up the mechanism by which the community selected its own corporation. It would also guarantee public access to channels.

The city's final and most expensive responsibility would be to lay the cable, "not as a gift" but as a loan to the profit-corporation that would pay back the cost, the same way you pay back a mortgage.

"The community would set up the nonprofit company and develop the capacity to monitor the operation." It would also recruit employees for the system, aid in the establishment of indigenous business firms to service the system, and supervise the promotion of community residents into various management positions. Atkins claimed that these functions of a nonprofit community corporation would prepare the community for complete ownership.

The private operator "would provide the technical know-how and the capital to set up the operation." A large amount of the cost would already be absorbed by the city under Atkins' plan, and the corporation would only have to pay back the cost of the cable.

Incentives for private corporations to undertake such a venture are the ability to "sell tax depreciation which the cable system would produce in large quantities" and the fact that much of the initial cost would have been absorbed by the city.

"Whether we wind up with a black-owned company as the general partner in this joint venture, or a white-owned company, the community is there," Atkins explained.

This plan followed an earlier premise that the best means of securing community control was through what Atkins called a transfer done in "an orderly fashion."

"By what process do we acquire community control, or work toward community control?

"Even if you accept the argument that in many instances, if not most, communities are not equipped to control, can't afford to control or aren't ready to control the system, that doesn't mean we'll never be ready. So maybe we ought to be talking about the franchise grant itself, a process by which control is transferred in an orderly fashion.

"It might be five years; it might be ten years; it might be fifteen. In some instances, I suspect communities would even be willing to wait twenty years if they knew they were going to get control in an orderly way," Atkins declared.

Part of his own effort to protect community interests was an ordinance he introduced to the Boston city council two years ago requiring a local referendum approval of any CATV franchise.

He has also been a strong opponent to franchises awarded by uninformed local governments. He pointed out that ignorance about cable leads to the neglect of the black community's interests.

On the issue of cable system construction, Atkins said that whoever owns the system—a black corporation or a white corporation—cable has to be laid and someone will get those jobs.

"Are we going to talk about having a cable television system built in our community the way everything else is built in our community? Where people from our community have damn little to say about it, and a hell of a time getting a job doing it? No. We don't even clean our own streets. So maybe it's visionary for me to think that we might string our own cable. But I think it's time to ask that question."

---

ownership and management

PANELISTS:

MORTON JANKLOW, Attorney, Janklow & Traum, and CATV Consultant to the Bedford-Stuyvesant Restoration Corp., New York, New York.

PAUL KAGAN, Paul Kagan Associates, CATV Financial Consultants, New York, New York.

TED LEDBETTER, Engineer and President, Urban Communications Group, CATV Management and Engineering Consultants; Washington, D.C.

SAM STREET, President, S. S. Street Associates, CATV Consultants, Washington, D.C.

FRANKLIN THOMAS, Attorney and President, Bedford-Stuyvesant Restoration Corporation, Brooklyn, New York.

On the third day of the workshop, panel discussions turned to "the nuts and bolts" of setting up cable system ownership and management.

Sam Street, President of S. S. Street Associates and a cable owner for 10 years, opened the discussion with a pointed description of what is needed to set up and run a cable company.

"After you get the franchise, you have a piece of paper in your hand from the city which says you can go into the CATV business," Street began.

"This is a legal document, and it gives you the right to construct. You have this and that's wonderful," he said, "but beginning the construction is still a long way off."

The first obstacle is the telephone company. "If you are going to put your cable on telephone poles, the telephone company must approve an application, The telephone company gives you a pole attachment contract which is probably 20 or 30 pages long asking where you're going to put your cable, and most important, who you are, and do you have the money to put up this piece of cable on their telephone poles.

"They are going to charge you anywhere from $3 to $5 per pole per year," he continued, "and that adds up to a lot of money when you're talking about a city like Washington or Detroit. It might run $50,000 to $70,000 per year. You cannot construct one foot of cable until you have a signed contract from the telephone company."

Before the pole attachment is signed, the applicant must give the telephone company a list of stockholders, a performance bond costing $15,000 to $25,000 and an insurance certificate that protects the telephone company from liability in case of an injury during construction and operation.

A similar contract has to be signed with the power company.

There are conflicts between the telephone company and CATV operators over what can be put on the cable that is running on "their telephone poles," Street said, and this conflict has not yet been resolved.

Moving to the second step, Street said, "Let's consider that you get all your insurance certificates, get your performance bonds, you have your pole attachment contract for the power company and for the telephone company.

"The next step is indicating to the telephone company which poles are going to be used so that each one can be inspected.

"Every one of those poles that you're going to be on has to be checked by a telephone man; this is time-consuming.

"Every pole has to be certified. It has to be written down that you're using it, because again, you're going to be paying rent on it." Pointing to what he called "hidden costs in operating a cable company that most people don't realize or fail to acknowledge," he said a pole at intersections congested with other wires has to be serviced. The wires will have to be rearranged to accommodate the cable. He estimated the charge for this service is between $50 to $200 per pole.

The third preliminary step in actual construction is applying to the FCC for the right to carry broadcast TV signals from another area.

Attorney and Consultants

"I strongly recommend the first thing you do is hire an FCC attorney in Washington, and that's a must. You can make some very expensive mistakes if you don't have an attorney and his job is to make sure that everything is legal and that things will go through relatively smooth.

"The FCC has to approve your application before you can go into business," Street said, because the FCC has to approve the use of over-the-air signals from another state. The FCC will also determine how many channels can be carried over the system.

After FCC approval is secured and all contracts signed, the next phase is construction.

"First thing I would do would be to hire a CATV consultant," Street said. "We're not really in competition with each other. We're in competition in getting a franchise, but once we get into operation, we're not going to come into Washington if you're already here or you're in Detroit. Cable franchises are relatively exclusive, and a lot of people are very helpful in this industry."

With the help of a consultant, preferably an engineer, the new company must decide what kind of system to build. There are five or six different kinds of choose from, Street said, and the choice will determine "how much money you are going to spend, when you will get a return, and how you can expand." The choices Street named are dual cable, two trunks, single trunk with converter, and a re-diffusion telephone-type switch system. That's where an engineering consultant is needed.

To get the system constructed, Street suggested that a new cable company "go turn-key." Under a turn-key operation, a construction company (usually a hardware and equipment manufacturer) is hired by the cable company under contract to build the system. The construction company is paid by the cable company and after the system is built, "turns the key" over to the cable operator.

"I would, for a new company, go that way," Street advised. "I feel that there're too many costly errors to try to build it yourself. I would stipulate in the contract that you want to use local labor, and this is where the community input comes in. You want to train local people, because they're going to be your technicians and installers. These are details that you can work out with the manufacturer.

"But I think you ought to use a combination of a consultant and a turn-key construction company."

Calling financing "probably the hardest part," Street said he would leave that topic for later and treat the next problem, which is paying for the cable.

$5,500 per mile

"We've done cost analyses of some suburban metropolitan areas in which we're building a 24-channel system and it costs $5,500 a mile. That's our estimate and we think it's rather accurate.

"However, our engineer tells us, if we go underground, look for $20,000 to $25,000 a mile and that just blows your projections out the window. So when you get into a place like Washington and it's all underground, you're looking at $20,000 to $25,000 a mile in most areas. And that's rather frightening."

Using above-ground cable, the cost estimate is $5,500 per mile. Street called aerial cable "the best way to go." The worst way, he said was to pay the telephone company to build the system. This option, called a "lease-back," makes the cable company rent the system from the telephone company for ten years.

"Lease-backs are notorious money-losers for the operator, and you end up working for the telephone company and being a sales agent without the control over the system or resale value of that system. When you're all finished paying for it," Street continued, "you don't own it, and you have no equity in it."

Street said that while the system is being constructed, the cable company must begin hiring staff, because "when your first hook-up is made, you are in business.

"Your key people are, first, your manager, who is in charge of the office-promotion and marketing, and second, your chief technician or system engineer.

"Your system manager is in charge of all your clerical help, and he's over-all manager, and your system engineer is in charge of your installers and your technical people."

Thinking Like a Businessman

Paul Kagan, still talking ownership and management, explored the possibilities for selling cable television to big city subscribers where there are few problems of bad reception. The primary selling point will be additional channels and the ability to show customers the kind of programming they want—like hometown sports.

"In the big city it is harder to saturate a CATV system than it is in a small or medium-sized city, and if it's harder, then it must be more expensive, and if it's more expensive, then you'd better start thinking like a businessman, or you're going to get very, very frustrated, And frustration you don't have to buy at this point."

In a city of 350,000 persons and 100,000 homes, given all the equipment that must be bought, Kagan said the cost would be over $10,000 per mile. In a city of 350,000, the system would be 700 miles long for a cost of over $7 million.

"In Akron, they're building a 1,100 mile system, and their investment is already $15 million," he related.

Unlike cable operators who started in mountainous rural areas where reception is extremely poor, the cable owner in a big city will have a selling problem.

"In a big city, you may only be able to sell 10 or 15 percent in the first year or two," he cautioned. "If you're a cable nut, and if you're a good businessman too, you may try to figure out ways to get people to subscribe, to make the business viable."

He cited the example of cable companies in the New York area which offer their subscribers televised hockey and local basketball games by paying for the television rights. He added that many cable operators want to show first-run movies for an extra charge to help make enough money to pay for the system. "Five dollars a month (service charge to customers) for straight television service isn't going to pay off that investment fast enough for you to be able to raise the money to build the system."

Cable operators will have to have a high return on investment in order to convince potential financers to put up the estimated $7 million needed to build the system. Therefore, soliciting CATV city subscribers becomes all important.

Four sources of traditional financing Kagan outlined were insurance companies, loans, the sale of public stock, or the formation of a limited partnership. A fifth source, and the one he recommended for cable companies working in the center city, was "a public offering of limited partnership interests."

Insurance companies were the first sources of financing for cable companies. When the industry began to grow, insurance companies asked for so much of the stock in return for the loan that CATV operators began to look elsewhere for financing.

"Money is still coming from the insurance companies, but as a percentage of the total amount of money that's going to be raised for CATV, it's getting smaller," Kagan said.

"It is pretty hard, if not impossible, to do a public offering (of stock) when you're just starting up and don't have any experience or any systems in your company," Kagan observed, adding that most companies that sell public stock are already established and have "a track record" including profits. "The odds are that a public stock offering of the conventional type, of common stock, is probably not in the cards for you for getting your initial capital."

The limited partnership route to financing has attracted many cable operators. The operator's limited partners are usually "people in high tax brackets who need tax losses to offset income," Kagan said. The advantages are "you get to own and operate the system, he (the investor) gets to finance it, and gets the benefit of the losses during the early years of the business when the company is not breaking even."

When the company begins to make money, "the full ownership can revert back to you, and you can provide some stock in your company to the original investor, and he ends up with a long-term capital gain way down the road." The conventional way to find limited partners is to seek people with high incomes, possibly "a famous name" (a technique used by TelePrompter) to put up the money for the system.

The fifth area of financing and the one favored by Kagan for cable operators in minority communities, is the offering of limited partnership interests to the public. An example of this type of venture is LVO Cable in California. LVO "sold 500 people units (shares), and through that raised two and a quarter million dollars. This was the first project. It would have to be labeled the pilot project."

In his concluding remarks, Kagan observed that white companies are considered sound investments, but do not have the political power of community organizations. On the other hand, black businessmen must convince the financial community "that you're good business people, and that's why you need to understand what the problems are."

Most important, he said, was being able to put together a prospectus that shows a maximum number of subscribers in the market area being hooked up to cable "because that's the only way that investment will ever get paid off."

Ending on an optimistic note, Kagan said "when you have the nuts and bolts figured out, and you put out a prospectus, knowing the community involvement you already have, you're going to have a very interesting story to tell about selling communications in the central city, and that's the concept I think you should be following."

Money For The Community

Getting money to finance cable television through organizations like the Bedford-Stuyvesant Restoration Corporation offers a unique opportunity because of the political leverage of an organized community.

That was the main theme of the last two panelists, Franklin Thomas, President, and Morton Janklow, Special Counsel, for the Bedford-Stuyvesant Restoration Corporation. Traditional financing methods won't work for fledgling cable companies, especially in the inner city, Janklow declared. Janklow founded the San Diego cable system which has grown to be one of the largest in the country. He has since been assisting the community development effort in Bedford-Stuyvesant, one of the nation's most economically depressed areas.

"We at Bedford-Stuyvesant have a theory that this business can make money and that you can attract partners who are interested in making money and not necessarily in doing good," he explained. A joint venture between a community development group and a profit oriented corporation can work in the black inner city, he said.

One incentive for the corporation to join the community group is that the corporation may not be able to get a cable system going without community support, Janklow observed. Another may be that the corporation can market products in the community. The kinds of businesses that may be attracted to such a venture are "major corporations which are either interested in the technological features of the industry because they manufacture equipment, or they manufacture sets, or they manufacture peripheral devices which are going to be significant," Janklow pointed out.

Under the type of joint venture described by Janklow, during the first years of the cable company's life, the corporation that goes into the partnership with the community group would make all the profits.

"Defer your participation in profits until such time as your money partner has his investment out—however long it takes, seven or eight years," he advised.

According to Janklow's idea, "during those seven or eight years" when profits and losses are accrued by the "money partner," the community group should develop training programs, employment potential, community economic growth potential and "be creating a financial package which can be very profitable down the pike."

This kind of partnership is not now in effect in Bedford-Stuyvesant but Janklow predicted that "we can analyze the cost to a point where we can discuss with a corporate partner what would be in it for him."

Community Advantage

He said a company that wants to establish cable television in a black community would find it an advantage to have a community group as a partner because that group's opposition could kill the company's plans politically. He added that it politically benefits a city government to grant franchises to community organizations rather than an outside corporation.

"There's a tremendous political implication to the Mayor of the City of New York if a community-sponsored group with financial backing begins to develop the use of cable, not only for profit and not only to generate revenue, but also to have an impact on the community," he noted.

And if a community group is in partnership with a profit-making corporation, foundations will be more likely to offer further financing.

On the issue of owning the entire system, as opposed to merely having access to the system, Janklow said, "people ask, why Bed-Sty doesn't just go in and muscle the city and demand two of the channels that's going to be for their use? The answer is, that's not what we're after. We could do that very easily. What we're after is community development—profits plowed back into the community—a kind of militancy and a kind of use of the system that won't be very acceptable to somebody else."

In the United States, Janklow explained, "control rests with the owner to a great degree. That's why you want to be the owner."

When someone else owns the system they can determine what time and for how long programs will run on any of the channels. If a community group owns the station, it can make those decisions.

"My last word to you is, start the job—start the ball rolling now."

Ownership Or Access

Franklin Thomas, President of the Bedford-Stuyvesant Restoration Corporation, followed Janklow with a description of how minority community organizations are traditionally encouraged to gain access rather than ownership of the system. Arguing for ownership because of the business opportunity as well as the potential political and cultural power, Thomas pointed out that of all industries, cable television is "uniquely one that you think would fit right into all of the talk about local economic development."

"It's a relatively contained industry. The major source of revenue is subscriber fees. The subscribers in your case would be the residents in your own community. The technology has been reasonably standardized—you would think that here's an opportunity to perhaps build in a system, in a business, which has as its base the very community that you're trying to serve."

Thomas said that the political power of a community group can help offset opposition from the telephone company in granting permission for use of their poles to string cable. He added that a strong community corporation can make it politically unfeasible for a local government to pass over it and award a franchise to a "strictly commercial" corporation.

Among the questions raised during the audience exchange was what can be done to meet bonding requirements on construction. One panelist suggested that beginning cable companies should use the turn-key method with the contracted construction company responsible for posting bond. In most cases, these companies are not required to put up a performance bond because they have established reputations.

Audience reaction to joint-venture financing was mixed. There was also a strong reaction against sub-contracting (turn-key) construction.

The crux of the objection to sub-contracting was that if an outside company builds the system, the community remains ignorant of cable technology. One member of the audience said, "you always end up sub-contracting. The power is in the knowledge and the technology." He added that community groups need to know the ABC's internally.

"There must be a designated program saying that this (the cable system) is totally inclusive and for the good of black people. . .for the good of the black community."

Janklow reacted strongly to the negative expressions by the audience on these issues. "I don't think we ought to come away with this concern about bonding problems." He added that corporation incentives like bonding requirements and tax breaks are best settled by an attorney.

"You pay tax lawyers to worry about that. That's not the critical issue. The critical issue is how, given the need to do this and the desire to do it, how to raise the money."

He suggested that methods of raising money can be narrowed to four areas: federal agencies, foundation loans and grants, partnership with large corporations and wealthy individuals.

With a foundation grant of $1 million, a community organization may be able to persuade a bank to loan it an additional $3 million, the panelist suggested. He said banks are developing a "sense of public spiritedness" and many feel that such a loan is "good public relations."

Reflecting the optimism that prevailed among most of the workshop participants, one panelist said, "It would be a great mistake to make the issue so complex that you lose sight of the goal."

The joint-venture partnership plan raised the question of who is in control and who would reap most of the benefits from the system—the community group or the big corporation. The panel response was, "You seem to have some paranoia about talking about limited partnership and joint ventures. I think that while everybody here is in financing, we just simply need to remember that in joint ventures, certain questions are negotiated; particularly in the area of profit . . . and (also) in terms of who's in control."

Citing the fact that all considerations about financing and operation must be carefully weighed, Ledbetter concluded,

"We're all going to go and do more work on it and we're going to learn more on it all the time. Community development corporations and the other technical assistance groups here are experienced in financing and what we've tried to do with this workshop is to plug in the information that is specific about cable television; information that makes it easier to determine what makes cable different or what makes cable the same as other kinds of business management and financing. I hope that's what we are accomplishing here today."

"communications is power"

Nicholas Johnson

Cable television "is going to be a source of employment. But I think it's a mistake to overemphasize that. The thing you should really keep your focus on is communications as power.

"It's economic opportunity, it's jobs, it's lots of good things. But none of those compare to what it provides in terms of social, political, and economic power in this country."

With this analysis, Nicholas Johnson, outspoken member of the FCC, delivered the workshop's final keynote address. Johnson broke with the traditions of FCC members by strongly opposing the present practices of the broadcasting industry which he says has misused its power. He said that cable television offers a chance to harness the power of communications for the good of the community and to alter the present state of mass media.

"Information is power. People will act if they are informed. And that's what led one television commentator to say that the act of getting 100 million Americans to focus their attention in the prime time evening hours on westerns and situation comedies is a political achievement, the equivalent of the Roman circuses. A political achievement. Now, you've got to remember that. This is no accident."

Johnson explained that television was originally viewed as an entertainment medium with little public responsibility. Now the same people who own newspapers and magazines, own TV. Politicians who want to buy time on television or space in newspapers have been turned down as have groups who opposed government policy.

Two examples he gave were the Democratic congressmen who wanted to buy time to answer the President's State of the Union address and a group of white businessmen who wanted to speak out against the war in Vietnam. Both were turned down by television stations.

"We had some businessmen who decided that the war was bad for business," Johnson related. "So they came to WTOP in Washington with money in their hand and thought that WTOP had time to sell them. They walked in and said, 'Here's our money. We'd like to buy a minute or two on your station to run this message about the war and how it's bad for business.' And WTOP said, 'No, we aren't going to sell you any time.' Well, if you want to see an instant radicalization of some businessmen, there it was because here they're used to 'money buys everything.'

"Then there were fourteen United States Senators who figured that since the President of the United States got free time on all three networks in prime time in the evening, they ought to be able to buy a half hour on one network in some bad time. So these 14 United States Senators went in with their money and said, 'We'd like to buy some time to answer the President of the United States because he's talking about an issue—the Southeast Asian war—that's in the interest of some Americans. And we think there ought to be some other point of view expressed in our society besides just this one point of view. And we've been elected officials in the United States Senate, and therefore, we'd like this time.' The network said, 'No, you can't buy it. We aren't going to sell it to you.'

"Well, the point is made. If white businessmen can't buy time and if United States Senators can't buy time, you're damn sure not going to get it for free."

Cable television has the potential to open up the communications industry because it eliminates the problems often referred to by broadcast television stations. There is no channel scarcity and there's still room for community groups and especially black people who have been eliminated from any power in mass media to get some control. Johnson referred to half inch video equipment that children can be trained to use. Tapes made with this equipment can be easily shown on cable television. Johnson said this video equipment costs as little as $1,500.

"They used to say 'the pen is mightier than the sword' in terms of revolutions. And today that pen is represented in that video camera. That's power. That really is power.

"You want as many young black people in this country as possible to have firsthand experience working with that equipment while they're young. That's important. It's terribly important."

The simple operation of this type of camera will allow many people to make films that can be shown on cable.

Cable television, because of its large channel capacity, can also provide educational programs from preschool through college. It would be possible for people to earn their high school diplomas by watching television at night.

"You can put all kinds of other stuff on. You could have practical information on health care. You could have a channel, for example, that just had a constant 'crawl' on jobs available in the city. What the job is, what it pays, what telephone number you call. And any man or woman sitting at home watching television could just turn to that channel and see what kinds of jobs are available.

"All kinds of information. Information is power," Johnson repeated. "Information is locked up. I mean that's how you measure power today. More than armies or money or anything else it is access to information and control of the mass communications. The president knows that. Agnew knows that. That's what all that's about."

Community development groups should work to get ownership of cable companies but they should also make sure that they offer community service, and keep the rates as low as possible. In the area of ordinances, he advised that material be submitted to the city for referendum approval, that it allow for as high as 40 channels, and that channels be set aside for educational purposes, especially use by the schools, use by the local government, and use by private citizens.

Concluding that the FCC cannot be relied on to clear the way for community ownership or for progress in programming for cable, Johnson advised that the best arena for involvement is at the local government level.

"You've got to have a political base and you've got to have some power, something that you can point to and rely on and use as a base when you go talk to the guys that are running the town so that they realize that there is a reason why they have to listen to you," Johnson concluded.

Johnson's term as an FCC Commissioner ends in 1973. He was appointed by President Lyndon Johnson in 1966 to the seven-year position. There are not now, nor have there ever been, any black members of the FCC.

Nicholas Johnson said he attempted to get a black person appointed, but was told that efforts to find an "acceptable" black member have been unsuccessful.

speakers and resource persons

Thomas Atkins, City Councilman, Boston, Massachusetts

Don Bushnell, Acting Director of Watts Communications Bureau, Los Angeles, California

Gary Christensen, General Counsel, National Cable Television Association, Washington, D.C.

Cliff Frazier, Director, Community Film Workshop Council, New York, New York

Stan Gerendasy, Public Broadcasting Consultant to the Ford Foundation, New York, New York

Vernard Gray, Director of Fides House, Communications Workshop, Washington, D.C.

Ellis Haizlip, Executive Producer for "SOUL" WNET-TV New York, New York

Cliff Henry, Deputy Director of the National Council for Equal Business Opportunity, Washington, D.C.

M. Carl Holman, President of the National Urban Coalition

James Hudson, Partner in the firm of Hudson and Leftwich, Attorneys at Law, Washington, D.C.

Morton Janklow, Partner in the firm of Janklow and Traum; Attorneys at Law, New York, New York

Nicholas Johnson, Member of the Federal Communications Commission, Washington, D.C.

Paul Kagan, President of Paul Kagan Associates, CATV Consultants, New York, New York

Bertram Lee, Community Development Consultant, Boston, Massachusetts

John Lenear, President of Metro Publishing Company, Cleveland, Ohio

Roy Lewis, Cinematographer and Assistant Director of Project Reach, Notre Dame University, Notre Dame, Indiana

Edward Lloyd, President of West Essex Cable TV Company,
Newark, New Jersey

Jack McCarthy, Vice President of United Utilities, Inc., Washington, D.C. and Kansas City, Missouri

Bernard McDonald, Director of Comprehensive Manpower, Bedford-Stuyvesant Restoration Corporation, Brooklyn, New York

Gil Mendelson, Research Director, Unity House, Washington, D.C.

Early Monroe, Engineer, CATV Bureau, Federal Communications Commission, Washington, D.C.

Geoffrey Nathanson, President, Optical Systems Corporation, Los Angeles, California

Arthur Peltz, Communications Specialist for the Community Relations Service of the Department of Justice, Washington, D.C.

Steven Rivkin, Attorney, Washington, D.C.

Sol Schildhause, CATV Bureau Chief, Federal Communications Commission, Washington, D.C.

Ralph Lee Smith, Sloan Cable Commission, New York, New York

Sam Street, President of S. S. Street Associates, CATV Consultants, Washington, D.C.

Stuart Sucherman, Program Officer, Ford Foundation, New York, New York

Franklin Thomas, President of Bedford-Stuyvesant Restoration Corporation, Brooklyn, New York

Phil Watson, Coordinator for Broadcast Development, Howard University, Washington, D.C.

Morissa Young, Writer/Researcher, Black Efforts for Soul in Television, Washington, D.C.

participants

Afro-Urban Institute
2210 North 3rd Street
Suite 310
Milwaukee, Wisconsin 53212
Jim Estes
Larry Reed
Henry Crosby

Bedford-Stuyvesant Restoration Corporation
1368–9 Fulton Street
Brooklyn, New York 11216
Franklin Thomas – President
Barry Lemieux
Bernard McDonald

The Black Congressional Caucus
The U.S. House of Representatives
Washington, D.C. 20515
Carolyn Jefferson

Black Development Foundation
1308 Jefferson Avenue
Buffalo, New York 14208
Diane Anderson

Black Economic Union of Greater Kansas City
2502 Prospect
Kansas City, Missouri 64127
Curtis R. McClinton, Jr. – President

Black Strategy Center
8623 South Avalon
Chicago, Illinois 60619
Reverend C. T. Vivian
Al Sampson

Black United Front
2237 Georgia Avenue, N.W.
Washington, D.C. 20001
Absalom Jordon
Reverend Douglas Moore

CATV Marketing, Inc.
36 Quail Court, Suite 4
Walnut Creek, California 94596
Michael Berkowitz

Chicago Defender
1712-16th Street, N.W.
Washington, D.C. 20009
Ethel Payne

Community Communications Systems, Inc.
2516 Edgecombe Circle, North
Apartment 202
Baltimore, Maryland 21215
Lester Green
John Green

Cooperative Assistance Fund
1325 Massachusetts Avenue, N.W.
Suite 303
Washington, D.C. 20005
Edward C. Sylvester, Jr. – President

Cummins Engine Foundation
1001 Connecticut Avenue, N.W.
Washington, D.C. 20036
Ivanhoe Donalson

Dayton Youth Movement
1451 West 3rd Street
Dayton, Ohio 45408
Floyd Johnson

Drum and Spear Press
1371 Fairmont Street, N.W.
Washington, D.C. 20009
Courtland Cox

Federal Communications Commission
1919 M Street, N.W.
Washington, D.C. 20036
Doris Cole

FIGHT-ON Corporation
393 Clarrissa Street
Rochester, New York 14605
DeLeon McEwen, Jr.
Chairman and President

Foundation for Community Development
P.O. Box 647
Durham, North Carolina 28202
Adolf Reed
Pauline Bowman

Greater Memphis Urban Development Corporation
P.O. Box 224
Memphis, Tennessee 38101
Maeola Killebrew
Joe Purdy
Fred Hooks

Harbridge House, Inc.
11 Arlington Street
Boston, Massachusetts 02116
Harold Tincher
James R. Haynes

Harlem Commonwealth Council
306 Lennox Avenue
New York, New York 10027
James Dowdy – Director
Marygale Lasher
Armond Gilbert
Barbara Norris
Yvonne Thomas

Hough Area Development Corporation
7016 Euclid Avenue
Suite 336
Cleveland, Ohio 44103
Franklin R. Anderson – Director
Mrs. Archie B. Lewis

Inner-City Business Improvement Forum
6072-14th Street
Detroit, Michigan 48208
Peggy Fulton
Walter McMurty, Jr. – Director

McKissick Enterprise
360 West 125th Street
Suite 8
New York, New York 10027
Floyd McKissick – President

Model Cities Planning Council
Dayton, Ohio 45401
Silas Cox

National Bankers Association
1325 Massachusetts Avenue, N.W.
Washington, D.C. 20005
Bob Davis
Ron Hill

National Business League
4324 Georgia Avenue, N.W.
Washington, D.C. 20011
Berkeley Burrell – President

National Cable Television Association, Inc.
918-16th Street, N.W.
Washington, D.C. 20006
Wally Briscoe

National Council for Equal Business Opportunity
1211 Connecticut Avenue, N.W.
Suite 310
Washington, D.C. 20036
Benjamin Goldstein – Director
Samuel Daniels
Arthur Sparrow
Chris Powell
Charles E. McGriff

National Progress Association for Economic Development
100 West Coutler Street
Philadelphia, Pennsylvania 19144
Charles Taylor

National Urban Coalition
2100 M Street, N.W.
Washington, D.C. 20037
M. Carl Holman – President
Diane Kenny
Robert Hobson

Newark Cable TV, Inc.
287 Washington Street
Newark, New Jersey 07102
James Eddleton, Jr.
William Mercer

Office of Minority Business Enterprise
14th Street & Constitution Avenue, N.W.
Washington, D.C. 20235
Jay Leanse – Deputy Director
Paul Burke
George I. Washington
Vander Beatty

Opportunity Funding Corporation
1325 Massachusetts Avenue, N.W.
Washington, D.C. 20005
Jack Gloster – President
Paul Pryde

Peoples Development Corporation
1010 Vermont Avenue, N.W.
Suite 612
Washington, D.C. 20005
E. Richard Brown
Executive Director

Pepper, Hamilton & Scheetz Law Firm
Philadelphia, Pennsylvania

Sivart Mortgage Corporation
840 East 87th Street
Chicago, Illinois 60619
Dempsey J. Travis – President

Southwest Virginia Community
Development Fund
401 First Street, N.W.
Roanoke, Virginia 24016
Stanley R. Hale
Director of Administration
and Management

The Urban Institute
2100 M Street, N.W.
Washington, D.C. 20037
William Gorham – President
Tony Perot–Director,
Urban Fellows Program
Lacy Streeter

WTOP (TV and Radio)
40th and Brandywine Street, N.W.
Washington, D.C. 20016
Claude Mathews

Youth Organizations United, Inc.
912-6th Street, N.W.
Washington, D.C. 20001
C. M. Lewis

(1)

(2)

(3)

(4)

(5)

(6)

(7)

(8)

(9)

(10)

(11)

(12)

(13)

(14)

(15)

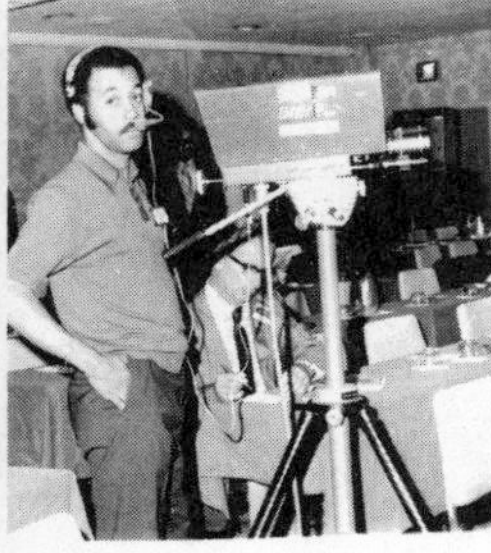

(16)

(17)

(18)

(19)

(20)

NOTES FOR PHOTOGRAPHS

From left to right

1. *Frank Anderson*
   Floyd McKissick
   C. T. Vivian
   DeLeon McEwen
2. *Roy Lewis*
   Jack Gloster
   Carl Holman
3. *Bill Wright*
   Early Monroe
   Phil Watson
   Frank Anderson
4. *Tom Atkins*
   Ben Goldstein
5. *Al Sampson*
   Tony Perot
   C. T. Vivian
6. *Tom Atkins*
   Carol Randolph
7. *Geoffrey Nathanson*
   Carol Randolph
   Bill Wright
   James Dowdy
8. *Panel presentation*
9. *Carol Randolph*
   William Perry
   Archie Lewis
   Ellis Haizlip
10. *Frank Thomas*
    Paul Kagan
    Ted Ledbetter
11. *Peggy Fulton*
    Cliff Frazier
    Gary Christensen
12. *Morton Janklow*
    Dempsey Travis
    Charles Tate
13. *Fran Rusan*
14. *Stuart Sucherman*
    James Hudson
15. *Kay Lindsay*
    Larry Reed
    Stuart Sucherman
    Henry Crosby
16. *Eric Stark*
17. *Morrissa Young*
18. *Nicholas Johnson*
    Carol Randolph
    Berkeley Burrell
19. *Gil Mendelson*
    (man in center with earphones)
20. *Dempsey Travis*
    John Lenear
    Cliff Frazier
    C. T. Vivian

SECTION 3

REFERENCE AND RESOURCE GUIDE

introduction

This section has been designed for use as a quick reference source for community development groups. The materials included are introductory, rather than conclusive or exhaustive.

References and citations presented identify a wide range of public and private organizations, and publications and periodicals that can provide comprehensive, detailed data on cable technology, franchising, system hardware, construction, financing, and programming. Many of these materials can be obtained without charge. A working library can be established for less than $200.

glossary of technical terms

AMPLIFIER—An electronic device used to boost a weak signal with minimal distortion.

ANTENNA SITE—The physical location selected for placement of the antenna tower. Hills, mountains and tall buildings are generally used as antenna sites.

CABLE TELEVISION (COMMUNITY ANTENNA TELEVISION) CATV—A method of distributing television signals through a wire rather than broadcasting those signals over the air.

CHANNEL—The segment of the RF spectrum to which a television station is assigned or to which a television camera is tuned when transmitting via radio frequencies.

CLOSED CIRCUIT—A system of transmitting TV signals to receiving equipment directly linked to the originating equipment by coaxial cable, microwave relay or telephone lines.

CONTRAST—The range of light and dark values in a picture, or the ratio between the maximum and minimum brightness values. For example, in a high-contrast picture there would be intense blacks and whites, whereas a lower-contrast picture would contain only various shades of grey.

COAXIAL CABLE—Copper or copper sheathed aluminum wire surrounded by an insulating layer of polyethylene foam. The insulating layer is covered with tubular shielding composed of tiny strands of braided copper wire or a seamless aluminum sheath and a protective outer skin. The wire and the shielding react electronically with each other and set up an electromagnetic field between them. This field reduces frequency loss and thus gives cable its great signal carrying capacity.

DISTANT SIGNALS—TV signals which originate at a point too far away to be picked up by ordinary TV reception equipment, or those signals defined by the FCC as outside a broadcaster's license area.

DISTORTION—The departure, during transmission or amplification, of the received signal waveform from that of the original transmitted waveform.

DISTRIBUTION PLANT—The hardware of a cable system—contains amplifiers and trunk cable, attached to utility poles or fed through underground conduits like telephone and electric wires.

DROP OUT—A black and white horizontal "blip" on the picture tube during playback of a videotape. Caused by missing video information. Common physical cause: missing iron oxide coating on videotape.

FRANCHISE—A legal contract between a city or county government and a company or corporation—containing terms and conditions for constructing and operating a cable facility in a specific political subdivision.

GHOST—A shadowy or weak image in the received picture, offset either to the right or left of the primary image. This is the result of transmission conditions which create secondary signals received earlier or later than the main primary signal.

FREE TELEVISION—A term applied to commercial over-the-air television. The viewer does not pay a direct charge for program production or signal reception—but does pay for such services indirectly through the purchase of consumer goods advertised over-the-air.

HEADEND—The electronic control center,

where incoming TV signals are amplified, filtered and converted to cable system channels.

HOUSE DROPS—The wire or cable that connects each building or home to the main (trunk) cable.

INTERCHANGEABILITY—The ability to exchange tapes between different videotape recorders with no appreciable degradation of playback image. Available at the present time only between machines of the same make and model.

INTERFERENCE—In a signal transmission path, extraneous energy which tends to interfere with the reception of the desired signals.

KINESCOPE RECORDING—A film recording made by a motion picture camera designed to photograph a television program directly off the front of a television tube. Sound is recorded simultaneously. Often called a "kine."

LOCAL ORIGINATION STATION—A studio equipped to produce cablecasts within the cable system operator's franchise area.

MICROWAVE—A method of transmitting closed circuit television signals through the air on a highly directional, line-of-sight system from the originating station to one or more receiving stations.

MONITOR—A special type of receiver used specifically in video reception, rather than radio frequency (RF). Video monitors are not tunable to channels, and are used for viewing tapes.

ORDINANCE—An official resolution by a local political body regulating an activity within the legally defined political subdivision.

PAY-TV—A system of charging viewers for an individual program or a special service. Programs and services may be transmitted over closed circuit or cable television systems.

PICTURE TUBE—A cathode ray tube used to produce an image by variation of the intensity of a scanning beam.

RECEIVER—A television set, designed for tuned (RF) channel reception of sound and picture. A receiver/monitor is a combination instrument capable of receiving RF or video and sending out video signals.

RECORDING HEAD (AUDIO)—A stationary assembly used to record or playback electrical impulses at audio frequencies.

RF—An abbreviation for Radio Frequency, a system of transmission utilizing tuned bandwidths of the radio spectrum to carry both audio and video signals as in commercial TV broadcasting.

SIGNAL—An electrical pulse. The electrical pulse of electrical energy. Signals are noted in terms of strength (voltage) and frequency (cycles per second). Audio signal frequencies range from 20-20,000 cycles per second; video, from 20 on up into the millions of cycles per second.

SPECIAL EFFECTS GENERATOR—A device permitting selection of several special combinations of images, supplied by one or more video inputs.

SUBSCRIBER—A person who purchases cable services. An installation charge and a monthly fee is paid for providing the connection to an audio-video distribution system.

SWITCHER—A control which permits the selection of one image from any of several cameras to be fed into the television display or recording system.

SWITCHER-FADER—A device permitting gradual, overlapping transition from the image of one camera to another. Sometimes incorporated as part of a special effects generator.

SYNC GENERATOR—A device used to supply a common or master sync signal to a system of several cameras. This insures that their scanning pulses will all be in phase. Scanning pulses out of phase produce distortion or rolling. This is sometimes noted as sync "loss."

SYNCHRONIZATION—The maintenance of one operation in step or "phase" with another. Abbreviated "sync."

TERMINALS—The connectors, transformers, and converters (if necessary) on the subscribers set.

TOP 100 MARKETS—Ranking of television broadcast areas by size of market, i.e. number of viewers and TV households. Used in FCC rulemaking and in selling of air time to advertisers in broadcast television.

TRANSFER—To go from videotape to film, or the other way around.

VIDEO—The visual components of a television signal.

VIDEO-TRANSMISSION—The picture signal applied directly to the viewing tube without use of an RF carrier frequency. As no conversion-reconversion stages are required, there is no picture deterioration. The result is a higher quality image.

VTR—Abbreviation for Videotape Recorder.

CABLE TELEVISION IN THE CITIES

facts and figures on catv industry

This section contains information, facts and figures about the CATV industry and operating systems.

Information Sources:

Television Digest—1970-1971 edition
Black Communicator—June 1971
National Cable Television Association Research Department

Growth of the CATV Industry

(as of January 1 of each year)

Year	Operating Systems	Total Subscribers
1952	70	14,000
1953	150	30,000
1954	300	65,000
1955	400	150,000
1956	450	300,000
1957	500	350,000
1958	525	450,000
1959	560	550,000
1960	640	650,000
1961	700	725,000
1962	800	850,000
1963	1,000	950,000
1964	1,200	1,085,000
1965	1,325	1,275,000
1966	1,570	1,575,000
1967	1,770	2,100,000
1968	2,000	2,800,000
1969	2,260	3,600,000
1970	2,350	4,500,000
*1970	2,570	5,300,000

**added by editor*

U.S. CATV Systems By Subscriber Size (As of Feb. 7, 1969)

Size by Subscribers	Systems
20,000 & over	8
10,000-19,999	50
5,000-9,999	144
3,500-4,999	123
2,000-3,499	279
1,000-1,999	423
500-999	427
50-499	730
49 & under	46
Not Available	260
Total	**2,490**

Media Ownership of CATV Systems

Of the 2,490 systems operating as of March 9, 1970, following is by media ownership:

Media	Systems	%
Broadcaster	910	36.5
Phone	146	5.8
Newspaper-publishing	207	8.2

Channel Capacity Of Existing CATV Systems (As of March 9, 1970)

Over 12	86
6-12	1,720
5 only	459
sub-5	61
Not available	164
Total	**2,490**

CATV ORIGINATIONS By Existing CATV Systems (As of March 9, 1970)

Current

Automatic originations		1,019
Time & weather	984	
News ticker	116	
Music	74	
Stock ticker	18	
Emergency warning	10	
Weather only	4	
Time only	1	
Test Patterns	1	
Daily program schedule	1	
Local originations		399
Local live	293	
Public service	51	
VTR	66	
Film	54	
Local news	13	
Educational	15	
Movies	9	
Local/Special events	4	
Local sports	4	
Advertising	4	
Local interviews	2	
Church Services	2	
Other	4	
TOTAL*		1,089

Planned

Automatic originations	226
Local originations	273
TOTAL*	416

*Figure isn't total of "automatic" & "local" originations due to overlap in categories.

TOP 25 CABLE TELEVISION COMPANIES

Includes all majority-owned subsidiaries

Company	Subscribers
1. TelePrompTer Corporation[1]	535,000
2. Cox Cable Communications, Inc.	197,000
3. American Television & Communications Corp.	180,000
4. Viacom[2]	150,000
5. Tele-Communications, Inc.[3]	141,590
6. Cypress Communications Corporation	140,000
7. Cablecom-General, Inc.	123,000
8. Time-Life Broadcast, Inc.	105,836
9. Jerrold Corporation	100,000
10. Midwest Video	90,000
11. Continental Transmission[4]	89,000
12. Television Communications Corp.	84,000
13. Service Electric Cable TV, Inc.	82,000
14. LVO Corporation	80,000
15. National Trans-Video, Inc.	75,000
16. Tower Communications[5]	65,000
17. Storer Broadcasting	55,000
18. Columbia Cable[6]	50,000
19. United Artists Theatre Circuit, Inc.	49,000
20. Continental Cablevision	46,500
21. General Electric Cablevision Corp.	45,000
22. Vikoa	42,000
23. Jefferson-Carolina	39,000
24. Philadelphia Bulletin	37,000
25. Rust Craft Cable Communications, Inc.[7]	35,000

(1) Includes the merger with H&B Communications Corp. and purchase of Reeves Telecom's systems.
(2) Spin off from CBS
(3) Includes merger with Centre Video
(4) Includes merger with National Communications Services.
(5) Includes merger of Citizens Financial Corp. and Communications Properties, Inc.
(6) Does not include the merger with Tele-Mark Communications.
(7) Formerly Neptune Broadcasting.

NUMBER OF SYSTEMS IN OPERATION PER STATE

State	Systems	State	Systems
ALABAMA	60	NEBRASKA	44
ALASKA	7	NEVADA	6
ARIZONA	22	NEW HAMPSHIRE	20
ARKANSAS	55	NEW JERSEY	26
CALIFORNIA	255	NEW MEXICO	26
COLORADO	29	NEW YORK	140
DELAWARE	8	NORTH CAROLINA	32
FLORIDA	76	NORTH DAKOTA	7
GEORGIA	54	OHIO	114
HAWAII	9	OKLAHOMA	55
IDAHO	45	OREGON	87
ILLINOIS	50	PENNSYLVANIA	296
INDIANA	40	RHODE ISLAND	1
IOWA	29	SOUTH CAROLINA	24
KANSAS	59	SOUTH DAKOTA	13
KENTUCKY	90	TENNESSEE	44
LOUISIANA	32	TEXAS	173
MAINE	25	UTAH	5
MARYLAND	26	VERMONT	25
MASSACHUSETTS	22	VIRGINIA	52
MICHIGAN	48	WASHINGTON	102
MINNESOTA	42	WEST VIRGINIA	129
MISSISSIPPI	50	WISCONSIN	46
MISSOURI	48	WYOMING	23
MONTANA	30		

TOTAL NUMBER OF SYSTEMS IN OPERATION 2,704

DISTRIBUTION OF CATV SYSTEMS BY NUMBER OF SUBSCRIBER

No. of Subscribers	*No. of Systems*	*No. of Subscribers*	*No. of Systems*
0-100	147	6001-6200	8
101-200	201	6201-6400	5
201-300	167	6401-6600	8
301-400	150	6601-6800	5
401-500	146	6801-7000	13
501-600	94	7001-7200	2
601-700	84	7201-7400	6
701-800	89	7401-7600	6
801-900	74	7601-7800	4
901-1000	82	7801-8000	6
1001-1100	40	8001-8200	3
1101-1200	49	8201-8400	5
1201-1300	45	8401-8600	5
1301-1400	45	8601-8800	4
1401-1500	59	8801-9000	5
1501-1600	41	9001-9200	1
1601-1700	28	9201-9400	
1701-1800	24	9401-9600	2
1801-1900	21	9601-9800	1
1901-2000	40	9801-10,000	4
2001-2100	38	10,001-10,200	2
2101-2200	22	10,201-10,400	4
2201-2300	15	10,401-10,600	2
2301-2400	23	10,601-10,800	1
2401-2500	20	10,801-11,000	3
2501-2600	14	11,001-11,200	2
2601-2700	10	11,201-11,500	1
2701-2800	16	11,801-12,000	2
2801-2900	13	12,001-12,200	5
2901-3000	26	12,301-12,600	2
3001-3200	24	12,801-13,000	3
3201-3400	18	13,001-13,200	
3401-3600	21	13,801-14,000	1
3601-3800	19	14,401-14,600	1
3801-4000	20	14,801-15,000	4
4001-4200	11	15,001-16,800	7
4201-4400	16	17,001-19,000	5
4401-4600	17	19,001-20,000	6
4601-4800	12	20,001-25,000	2
4801-5000	15	25,001-35,000	1
5001-5200	11	35,001-40,000	1
5201-5400	8		
5401-5600	7		
5601-5800	5		
5801-6000	11		

CITY—OWNED CATV SYSTEMS

Utilities Board-City of Opp
Box 311
Opp, Alabama 36467

Community Service Inc.
325 Ann St.
Frankfort, Kentucky 40601

Pitcairn Municipal Community
Antenna System
320 Center Ave.
Pitcairn, Penna. 15140

City of Crystal Falls
401 Superior Ave.
Crystal Falls, Michigan 49920

City of Norway Community TV
Main St.
Norway, Michigan 49870

Jackson Municipal TV System
504 2nd St.
Jackson, Minnesota 56143

Newburg Development Co.
Newburg, Missouri 65550

Town of Sumas TV Cable System
Box 5
Cherry St.
Sumas, Washington 98295

Village of Boaz
c/o Frank Stafford
Rt. 1
Muscoda, Wisconsin

GOVERNMENT OPERATED SYSTEMS

U.S. Army
Ft. Rucker, Alabama 36360

U.S. Govt. Directorate of
Maintenance
Electronic Branch
Ft. Leonard Wood, Mo. 65475

Ft. Sill TV Distribution
System
Signal Office Bldg.
Ft. Sill, Oklahoma

Information Officer
Perrin AFB
Sherman, Texas 75090

SUBSCRIBER-OWNED SYSTEMS

Bella Vista CATV Co.
Bella Vista, Arkansas 72712

Quincy Community TV Assn.
Box 834, Bradley St.
Quincy, California 95971

Canyon TV Service
Rt. #1
Cataldo, Idaho 83810

Mullan TV Co.
Box 615
202 2nd St.
Mullan, Idaho 83846

Pinehurst Co-op TV System
Main & Division
Pinehurst, Idaho 83850

West End TV Co-op Assn.
Box 264
Taylorville, Illinois 62568

SUBSCRIBER-OWNED SYSTEMS Cont'd

Lynch TV Inc.
Box 225
Lynch, Kentucky 40855

Ravenna TV Co-op
415 Elm
Ravenna, Kentucky 40472

Iron River Co-op Antenna Corp.
429 3rd St.
Iron River, Michigan 49935

Dufur TV Corp.
Dufur, Oregon 97021

Heppner TV Inc.
Box 587
289 N. Main
Heppner, Oregon 97836

Kinzua Corp.
Kinzua, Oregon 97849

Parkdale TV Cooperative
Rt. 1, Box 646
Parkdale, Oregon 97047

Blairs Mills TV Co-op
Blairs Mills, Penna. 17213

Williamson Road TV Co.
Box 145
Blossburg, Penna. 16912

Johnsonburg Community TV
Box 248
Johnsonburg, Penna. 15845

Millheim TV Transmission
Millheim, Penna. 16854

Port Clinton TV Cable Assn.
Main Street
Port Clinton, Penna. 19549

Rouseville TV Club Inc.
c/o Albert Brown
107 Myers St.
Rouseville, Penna. 16344

Marvin St. TV Assn.
902 W. King St.
Smethport, Penna. 16749

Irvindale TV Assn.
502 Jackson Ave.
Warren, Penna. 16365

Westfield Community Antenna
Assn. Inc.
Westfield, Penna. 16950

Tilbury Knob Inc.
Tilbury Terrace
West Naticoke, Penna. 18634

Youngsville TV Corp.
230 College St.
Youngsville, Penna. 16371

TV Assn. of Republic
Republic, Wash. 99166

Montgomery TV Association, Inc.
Box 607
Montgomery, West Virginia 25136

Rip-Shin TV Assn. Inc.
c/o Paul Green
104 Keyes Ave.
Philippi, W. Va. 26516

Pleasant View TV Cable
c/o Roy Valentine
Rt. 1
Rivesville, W. Va. 26588

Niagara Community TV Co-op
1509 River St.
Niagara, Wisconsin 54151

Woodman TV Cable System
c/o Henry Achenbach, President
Woodman, Wisconsin 53827

Largest U.S. CATV Systems
(Those with 10,000 & more subscribers as of March 15, 1971)

System	Subscribers
San Diego, Cal. (Mission Cable TV Inc.)	47,102
Northampton, Pa. (Twin County Trans-Video Inc.)	40,000
New York, N.Y. (TelePrompTer)	39,500
Allentown, Pa.	38,275
New York, N.Y. (Sterling Manhattan Cable TV)	28,212
Los Angeles, Cal. (Theta Cable of Cal.)	27,500
Harrisburg, Pa.	26,000
Altoona, Pa.	24,500
Santa Barbara, Cal.	23,500
Elmira, N.Y.	20,000
Utica, N.Y.	20,000
Williamsport, Pa.	20,000
Mahanoy City, Pa. (Service Electric Cable TV Inc.)	19,300
Cumberland, Md.	19,144
Johnstown, Pa.	19,103
Bakersfield, Cal. (Kern Cable Co. Inc.)	19,000
Eugene, Ore.	19,000
Easton, Pa.	19,000
Concord, Cal.	18,375
Atlantic City, N.J.	18,366
Santa Cruz, Cal.	18,000
Binghamton, N.Y.	18,000
Melbourne, Fla.	17,900
Toledo, Ohio	16,513
Harlingen, Texas	16,212
Parkersburg, West Va.	16,000
Palm Springs, Cal.	15,816
Everett, Wash.	15,684
Reading, Pa. (Berks TV Cable Co.)	15,620
Sarasota, Fla.	15,500
Macon, Ga.	15,400
Gainesville, Fla.	15,350
San Francisco, Cal. (TV Signal Corp.)	15,000
Tyler, Tex.	14,500
Martinez, Cal.	14,500
Lafayette, Ind.	14,500
Huntsville, Ala.	14,270
Florence, Ala.	14,000
Fayetteville, N.C.	13,500

Cont'd

Largest U.S. CATV Systems (Cont'd.)

System	Subscribers
Lafayette, Cal.	13,396
Colorado Springs, Colo.	13,075
Kalamazoo, Mich.	13,000
Lima, Ohio	13,000
Austin, Texas	12,900
Bakersfield, Cal. (Bakersfield Cable)	12,900
Dubuque, Iowa	12,900
Abilene, Texas	12,800
Charleston, W. Va.	12,500
Hazleton, Pa.	12,310
Rochester, Minn.	12,200
Ft. Walton Beach, Fla.	12,169
Ithaca, N.Y.	12,000
Palm Desert, Cal.	11,550
Ft. Smith, Ark.	11,500
Canton, Ohio	11,500
Pottsville, Pa. (Pottsville Trans-Video)	11,500
Clarksburg, W. Va. (American Cablevision Corp.)	11,500
Carmel-by-the-Sea, Cal.	11,148
San Jose, Cal. (San Jose Cable TV Service)	11,000
Flint, Mich. (Flint Cable TV)	11,000
Mansfield, Ohio	11,000
New York, New York (Comtel Inc.)	11,000
Pittsburg, Cal.	10,600
Danville, Ill.	10,500
Aberdeen, Wash.	10,500
York, Pa.	10,300
Seattle, Wash. (United Community Antenna)	10,300
Laredo, Texas	10,239
Marion, Ind.	10,100
Santa Maria, Cal.	10,000
Akron, Ohio	10,000

catv regulations and policy development

On August 5, 1971, the Federal Communications Commission presented a comprehensive package of proposed rules for regulation of cable television to the Office of the President and the Congress. The FCC indicated that it planned adoption action late this year with an effective date in early 1972.

From 1959 until June 1970, regulatory guidelines for CATV were established by the FCC and the Federal Courts primarily on a case-by-case basis. In June 1970, the FCC initiated a program to develop a comprehensive set of rules for cable regulation.

A synopsis of major regulatory actions and decisions during the period from April 1959 through August 1971 is presented here in chronological order for reference purposes. Lawyers and staffs of legal aid programs can obtain court decisions through established legal reference services. Copies of FCC documents can be obtained directly from the Commission.

Information Sources:

NCTA Research Department
Federal Communications Commission

MAJOR DECISIONS AND ACTIONS AFFECTING CATV: A CHRONOLOGY

April 14, 1959

FCC issued CATV and TV Repeater Services decision.
Commission found no basis for asserting jurisdiction or authority over CATV systems.
26 FCC 403 (1959)

February 16, 1962

FCC issued the Carter Mountain decision
Commission prohibited a microwave company from transmitting TV signals via microwave to a CATV system.
Carter Mountain Transmission Corp., 32 FCC 459 (1962)

July 15, 1964

Ninth Circuit Court issued Cable Vision decision.
Court held that CATV systems do not compete unfairly with broadcasters. (FCC's regulation of CATV is founded on the commission's view that CATV systems compete "unfairly" with broadcasters.)
Cable Vision, Inc. V. KUTV, Inc., 335 F. 2d 348 (9th Cir., 1964)

April 23, 1965	*FCC issued First Report and Order.* Commission asserted jurisdiction over microwave-fed CATV systems. 38 FCC 683; Memorandum Opinion and Order, 1 FCC 2d 524.
July 28, 1965	*FCC's Philadelphia Broadcasting decision.* Commission held that CATV systems are not common carriers. Later affirmed by D. C. Court of Appeals.
March 8, 1966	*FCC issued Second Report and Order.* Commission asserted jurisdiction over all CATV systems, and placed severe restrictions on carriage of distant signals in the top 100 markets. 2 FCC 2d 725; 47 C.F.R. 74.1101–1109 (1969)
May 23, 1966	*United Artists Television vs Fortnightly decision.* U.S. District Court ruled that a CATV system was liable for payment of copyright royalties. Decision later appealed to Supreme Court. 255 F. Supp. 177 (S.D.N.Y. 1966)
February 13, 1968	*FCC issued Community Antenna Relay Service (CARS) ruling.* Commission's rules and technical standards omitted provision for relay of programs originated by cable systems, thus barring CATV systems from interconnecting.
April 15, 1968	*U.S. Department of Justice Memo on CATV.* The Justice Department urged the FCC to allow CATV to develop as a competitive medium and to permit program originations and advertising.
June 10, 1968	*Supreme Court issued Southwestern decision.* Stemming from a case in which the FCC halted expansion of a San Diego, Calif., CATV system, the Court found the FCC to have such authority over CATV as "is restricted to that reasonably ancillary to the effective performance of the commission's various responsibilities for the regulation of broadcasting." United States v. Southwestern Cable Co., 392 U.S. 157, 160 nn. 19–22 (1968)
June 17, 1968	*Supreme Court issued Fortnightly decision.* Court, overturning a lower court ruling, found that CATV systems were not liable for payment of copyright royalties. Fortnightly Corp. v. United Artists Television, Inc., 392 U.S. 390 (1968)

June 26, 1968

FCC issued 214 decision.

After months of formal hearings, instituted at request of National Cable Television Association, the commission ruled that Section 214 of the Communications Act requires telephone companies to obtain FCC approval prior to undertaking the construction or extension of CATV lines.

General Telephone Company of California, 13 FCC 2d 448 (1968)

June 26, 1968

FCC began to assert jurisdiction over CARS originations and advertising.

In the Midwest Television decision, the commission approved CATV originations, but prohibited a CATV system from accepting advertising on its local origination channel.

November 18, 1968

U.S. District Court of Nevada ruled on TV Pix, Inc. vs. Taylor.

Court upheld a state law giving the Nevada Public Service Commission regulatory authority over CATV.

TV Pix, Inc. vs. Taylor, 304 F. Supp. 459 (D. Nev. 1968)

December 13, 1968

FCC issued Notice of Inquiry and Proposed Rulemaking.

Commission placed an almost immediate freeze on CATV. Through adoption of Interim Procedures, the FCC as a practical matter, placed its "proposals" into effect.

(Docket No. 18397) 15 FCC 2d 417 (1968)

December 13, 1968

FCC issued authorization of a nationwide broadcast pay-TV system.

Effective June 1969, the commission authorized a pay-TV system in those markets containing five or more grade A broadcast signals.

January 17, 1969

FCC issued first "clarification" of its Notice of Inquiry and Proposed Rulemaking.

Commission further restricted CATV systems' ability to obtain retransmission consent. Retransmission consent is equivalent to copyright clearance.

Final Report and Order 21 FCC 2d 307.

April 23, 1969

H.R. 10510-A CATV regulatory bill introduced in House of Representatives.

Rep. Samuel S. Stratton (D-N.Y.) introduced bill which would for the first time establish statutory guidelines for FCC regulation of CATV. Bill would grant FCC authority to require carriage and nonduplication of local signals by CATV systems, authority to impose reasonable technical standards and reporting requirements. Bill would also empower FCC to remedy situations where broadcasters could prove real—not imagined—economic injury resulting from CATV's operations.

May 16, 1969

FCC issued second "clarification" of Notice of Inquiry and Proposed Rulemaking.
"Clarification" appreciably increased the number of communities to which the new restrictions on CATV would apply.

May 24, 1969

Report of the President's Task Force on Telecommunications released.
Report recommended a relaxation of restrictive FCC policies on CATV, and declared CATV is the "most promising" method for expanding program diversity and localism.
President's Task Force on Communications Policy, Final Report.

May 28, 1969

National Cable Television Association Board of Directors approved the principles of National Association of Broadcasters/ National Cable Television Association staff compromise.
Compromise provided for payment of copyright fees by CATV systems, compulsory licenses for systems to carry all local signals, and exclusivity protection for broadcast stations.

Agreement also covered such areas as grandfathering, regulatory considerations and interconnections.

June 20, 1969

National Association of Broadcasters rejected NAB/NCTA staff compromise.

September 4, 1969

NCTA/NAB negotiations broken off.
NCTA broke off negotiations when NAB departed from original staff agreement and advanced new and more restrictive proposals.

September 12, 1969

FCC's Suburban Cable TV Co. decision.
Commission rejected a request that a factual test be undertaken to determine CATV's impact, if any, on broadcasting.

(Similarly, the FCC rejected another request to obtain empirical data on CATV in Valley Cablevision Corp., January 24, 1968.)

October 24, 1969

FCC issued rule requiring CATV origination.
Reversing its past position, commission permitted all CATV systems to originate their own programming. The FCC also required systems with over 3,500 subscribers to originate programming by April 1, 1971 and allowed all systems to carry advertising.
First Report and Order in Docket 18397

October 27, 1969

Supreme Court upheld FCC's 214 decision.
The court upheld the FCC's right to require telephone companies to obtain FCC approval prior to offering CATV service.

June 25, 1970

FCC issued a series of major rules and rulemaking actions in four areas affecting cable television.
Commission proposed a "Public Dividend Plan" which would allow CATV systems to import distant signals subject to specified payment to public broadcasting, deletion of distant commercials, and substitution of commercials of local UHF stations.

Second Report and Order in Docket No. 18397-A.
(1) *FCC Docket No. 18397-A Signal Importation*
Proposed rules concern the importation of distant signals from commercial and educational broadcast-television stations, revenue fees and copyright payments, and the use of channels within a single CATV system.
Commercial signals: A CATV system in one of the major television markets would be permitted to bring in four independent (non-network) commercial signals from outside areas. The cable system would substitute commercials from local stations for those on the distant signals. Preference would be given to commercials from local UHF stations, which then could sell advertising on the basis of larger audiences.
Educational signals: A CATV system in a major market could carry the signals of any number of distant noncommercial educational stations if the local educational licensee did not object. At the local educational licensee's request, the CATV operator would run appeals for funds for the local educational station.
Revenues: Major-market CATV systems which import signals from distant commercial or educational stations would pay 5% of their subscription revenues to the Corporation for Public Broadcasting, which would distribute one-half of its share to the local educational station and use the rest for other public broadcasting expenses.

Under these proposed procedures, the FCC says, Congress could—and should—pass legislation requiring fair compensation to program copyright owners.

CATV systems outside the Top 100 markets would not be affected by these proposals, except for any copyright payments called for by Congress.
Channel uses: All new CATV systems would be required to have sufficient channel capacity to provide one local origination channel; at least one channel for free use by local governments and for free political broadcasts during campaigns; public access channels for free use by local citizens and groups for presenting their views on matters of public concern; leased channels for commercial use by third parties; and instructional channels.

Systems with 20 or more channels would have to reserve at least half of them for these purposes.

In any case, remaining channels could be used to carry local and distant television broadcasts and specialized services.

FCC also proposed rules on cross-ownership, federal-state-local regulations, and technical standards.

(2) *FCC Docket No. 18891*: Ownership
Proposed rules contain the commission's proposals on cross-ownership of CATV systems and other media, and on multiple CATV ownership.

Cross-ownership: The FCC asked for comments on whether radio stations and CATV systems should be owned by the same person or company within a particular area, and, if so, to what extent. Comments were also invited on shared ownership, or shared use of facilities, between CATV systems and neighborhood or small-community weekly newspapers.

(FCC rules prohibit broadcast-television networks from ownership of CATV systems anywhere; they also ban ownership of a CATV system by a broadcast-television station in the same locality. The question of cross-ownership between daily newspapers and CATV systems is not considered in the FCC's current proposals; the commission is considering this issue in tandem with a previously proposed rule—Docket 18110—to ban cross-ownership between daily newspapers and broadcast-television stations and between daily newspapers and radio stations.)

The FCC also has asked for comments on whether cross-ownership should be allowed between CATV systems and CATV networks, microwave carriers, CATV-equipment manufacturers, national news magazines, advertising agencies and other entities whose ownership of CATV systems might not be in the public interest.

Multiple ownership: The commission has proposed that a single CATV owner be prohibited from having an interest in more than 50 CATV systems of 1,000 or more subscribers each. The limit would be 25 systems if the owner also had an interest in more than one television broadcast station, or more than two AM or FM radio stations, or more than two newspapers. Whatever the limit, the systems would have to be dispersed among markets of various sizes and among the states and multi-state regions.

As an alternative—or as a companion rule—the FCC has

proposed that a commonly owned group of CATV systems may serve no more than 2,000,000 subscribers, although subsequent expansion to 2,200,000 would be allowed.

(3) *FCC Docket No. 18892*: Regulations
The commission has declared three possible approaches: federal licensing of all CATV systems (much as broadcast stations are licensed now); federal regulations enforced through third-party complaints and through FCC hearings which could lead to cease-and-desist orders; and federal regulation of some aspects of CATV operation, with local regulation of others. The commission said it prefers the joint federal-local regulatory approach, with the commission establishing uniform or minimum standards for local regulators to follow.

Comments were invited on just what matters—such as rates, repair services and areas of service—should be of local concern and what, if any, the role of state regulatory agencies should be.

Franchise fees: The FCC proposed that franchise fees charged by local governments for a CATV system's right to operate could not exceed 2% of the system's gross revenues. Comments were asked on whether present fees—many of which are higher than 2%—should be allowed to stand.

(4) *FCC Docket No. 18894*: Standards
Proposed rules contain the commission's intentions to set standards that will insure good quality service to each CATV subscriber and require regular performance tests and reports. These standards may require most present CATV operators to readjust, and possibly to redesign, their systems.

Channel capacity: The FCC said it intends to specify a minimum capacity for a CATV system, which is based on the largest number of channels technically possible. It asked for comments on whether this should be 20 or 40 channels or some other figure, whether smaller capacities should be set as a minimum in smaller markets and how long a period should be allowed for present and new systems to comply.

Two-way system: The commission said it intended to require that future cable systems have a capability for two-way communication, so that at least a voice message return would be possible from a subscriber's home to a local program origination point.

Community channels: A cable system should supply a separate channel, available on a when-desired basis, for each distinct community within its franchised area, the FCC said.

June 25, 1970 cont'd

Deadline: The FCC suggested that all CATV systems be required to comply with all technical standards within three years of their publication.

August 17, 1970

Senate copyright subcommittee suspended efforts to obtain copyright revision (S. 543) in 91st Congress.
Senator John McClellan (D-Ark.) introduced legislation to extend current copyright bill. McClellan indicated he would introduce legislation containing CATV copyright provisions in the 92nd Congress.

December 8, 1970

H. R. 19926, a bill requiring the FCC to preempt CATV regulatory authority, introduced in House.
Cong. Robert O. Tiernan (D-R.I.) introduced bill giving FCC exclusive jurisdiction to regulate all aspects of CATV. Measure would allow FCC to delegate back to states and cities regulatory authority, if such was found to be in the public interest.

December 12, 1970

Justice Department criticized FCC's proposed CATV rules and urged commission to adopt a more liberal regulatory policy.
The department's antitrust division filed legal comments with the FCC urging the commission to allow CATV "to achieve its full competitive potential in an atmosphere free of overburdening obligations, conditions, and restrictions. . . ."

March 11, 1971

FCC opened two weeks of public hearings on its proposed new CATV rules. From March 11–25, 121 individuals representing public and private organizations and groups presented statements to the commission. Included were the National Association of Broadcasters, American Broadcasting Company, Columbia Broadcasting Station, League of Cities, Hughes Aircraft Corporation, Time-Life Broadcast, National Cable Television Association, Corporation for Public Broadcasting, National Education Association, Justice Department, Ford Foundation, Office of Economic Opportunity, National Football League and the Commissioner of Baseball. A number of minorities were also represented including the National Business League, Black Efforts for Soul in Television, Urban Communication Group, Equal Opportunities Committee of the Academy of Radio and Television Artists, and the Community Film Workshop Council.

March 11, 1971

NCTA Proposed a simplified seven-point CATV regulatory plan as an alternative to the public dividend plan. Among other things, the plan provided for four distant independent signals for CATV, public access to CATV channels, a failing station doctrine, expanded UHF nonduplication, and the existing "sports blackout."

May 27, 1971 *FCC suspends rule, First Report and Order in Docket 18397 requiring CATV systems with over 3,500 subscribers to originate their own programming.*

June 23, 1971 *President Nixon announces the establishment of a special White house committee to develop proposals for a comprehensive policy concerning cable television. Several cabinet members are appointed.*

July 15, 1971 *Senate Commerce Subcommittee on Communications holds hearing entitled "Community Antenna Television Problems" No. 92–12. Senator Pastore insists FCC 'check with him before finalizing rules and asked FCC Chairman Dean Burch when will they be ready.'*

July 22, 1971 *FCC submits preview of proposed regulations on CATV before the House Commerce Subcommittee on Communications.*

August 5, 1971 *FCC forwarded to Congress a 55-page explanation of its proposals but not final rules for regulating programming and other aspects of the CATV industry.*
The Commission said its objective in preparing the cable proposals was "to find a way of opening up cable's potential to serve the public without at the same time undermining the foundation of the existing over-the-air broadcast structure."

Final documents on the proposals will not be released until the "latter part of the year," the Commission said, in order to give Congress an opportunity to consider them. The Commission projected an effective date of March 1, 1972 for the new rules.

A summary of the major FCC proposals follows:

- The proposed rules, would require cable systems to carry all signals within 35 miles of the cable community or a minimum of three full network stations and three independent stations in the Top 50 market areas. Cable systems in other markets would be required to carry a minimum of three network signals and one independent.

- CATV systems will be required to offer the public one "access" channel. Anyone who agreed not to use obscenities, not to advertise or promote a "lottery" would have five free minutes.

- New CATV systems would be required to have a two-way capacity, allow signals to be sent from subscribers' home to a central monitoring station.

- CATV systems would be required to have a minimum capacity of 20 channels in each of the Top 100 markets.

- CATV operators would be required to make available a channel for five years at no cost to state and local governments and local educational groups.

- Although the Commission rejected licensing of cable systems, it outlined authorizing procedures. Prospective cable operators would be required to obtain from the Commission a "certificate of compliance" by submitting a form soon to be printed.

- CATV operators would also be required to show some machinery for handling complaints, approval of rates at a public hearing, a list of signals carried and an affidavit of service.

- Local franchising authorities would be required to set "reasonable deadlines" for construction and operation of cable systems to prevent franchises from remaining inoperative. The Commission said that the local franchising authority would also be required to place a "reasonable limit" on the length of a franchise. Pointing out that a franchise "in perpetuity" would be an "invitation to obsolescence," the Commission suggested 15 years as a maximum franchise term.

- The Commission said it would set technical standards to "assure the subscriber at least a minimum standard of reception quality, while at the same time permitting the continuation of technical experimentation." As cablecasting and other cable services are developed, technical standards will be applied in these areas.

- Discussing the matter of sports telecasts, the Commission noted that there are Congressional policies in this area (PL 87-331) and that cable systems will not be permitted to "circumvent" them. The Commission said "we intend to issue very shortly a notice of proposed rule making directed to this area, in order to ascertain the full thrust and purpose of 87-331 and how best we can formulate a rule to implement these purposes."

- Noting arguments that further regulation of cable should be delayed until copyright matters have been resolved, the Commission said that copyright policy is a matter for Congress and the courts and urged prompt action by Congress to enact a copyright law.

CABLE TELEVISION IN THE CITIES

special ordinance and franchise provisions

Special Ordinance and Franchise Provisions

City and county governments regulate cable systems by adopting an ordinance authorizing such regulation and by entering into a franchise agreement with a company or several companies to construct and operate a cable television system in defined geographical areas of the city or county.

Ordinance and franchise actions are the most crucial steps in the regulatory process. Community organizations should develop and submit specific provisions for inclusion in the ordinance that protect the interests of the community and prospective subscribers.

A summary listing and synopsis of suggested ordinance/franchise provisions that focus on minority issues and concerns are presented in this section for reference purposes. These suggested provisions have been drawn in part from the FCC Proposal Package submitted to the Congress on August 5, 1971; study recommendations of the District of Columbia City Council's Economic Development and Manpower Committee contained in the report "Cable Television in the District of Columbia," August 1971; Watts Communications Bureau "An Application to the City of Los Angeles for a Franchise to Install and Operate a Community Antenna Television System in the Los Angeles Basin"; a paper prepared by Jerrold Oppenheim of the Illinois Division, American Civil Liberties Union entitled "Soapbox Television: Model Code of Regulations—Cable Television—Broadband Communications," June 1971; and the material presented in Section I of this book by James Hudson.

Provisions recommended to the city council by community groups should be stated in legal terms. Since local situations and conditions vary greatly, a local decision must be made as to the practicability of including each suggested provision.

Suggested Provisions

General Requirements

Regulatory Body A separate agency or commission shall be established by the city to administer CATV regulations established by ordinance. The agency shall be composed of members appointed by the council or commission and shall include representatives of the various income, ethnic, racial, and religious groups who reside in the franchise area.

Franchise Fees Fees paid to the city by the franchisee shall be used to support the activities of the regulatory commission. Any surplus shall be allocated directly to minority organizations engaged in producing minority-oriented programs and to organizations engaged in training local residents for media occupations.

Length of Franchise The franchise should be effective for five years with a renewable option.

Franchise Areas In major population centers, uniform cable districts should be established to facilitate local ownership and control. A 20,000 household or subscriber system is economically viable in most areas. The largest single system now in operation has approximately 50,000 subscribers. Each franchise shall be geographically exclusive.

Prohibition Against Cross-Ownership Cable owners and operators are prohibited from having a controlling interest in or being under the control of other media owners, including newspapers, magazines, radio, motion picture, television, advertising firms, other cable systems, communications hardware manufacturing concerns, telephone and telegraph services, and repair or sale of television sets.

Conflict of Interest Elected and appointed government officials should not own any part of a franchise or serve in any official capacity with the franchisee organization.

Nontransferability of Franchise All franchises shall be nontransferable to discourage speculators who obtain franchises and hold them for future sale at a profit, with no intention of initiating construction. Nontransferability is also a means of assuring local ownership and control.

Channel Capacity A minimum of 24 channels shall be provided. Franchise applicants offering a greater number of channels at competitive subscriber rates shall be given priority consideration for franchise awards. Forty-eight, 54, and 60 channel systems are under construction in several cities.

Two-Way Capability The system shall have a bi-directional signal capability which shall be available to subscribers without extra charge. Audio, video and data transmission services from the subscribers' home to other points in the system, via an assigned channel, shall be provided.

Neighborhood Networks The system shall provide for branch networking that will enable subscribers in given neighborhoods to communicate with local schools, churches, municipal agencies without interference with the communications of other neighborhoods.

Interconnection Franchisees must interconnect their system with all contiguous cable systems and with any other system or service in such a way that subscribers can receive all channels of all interconnected systems and networks.

Subscriber Fees Fees will be regulated according to standards set by the city through the CATV Commission. Rates must insure no more than a fair return on original cost less depreciation of the system.

Installation and Hookup Charges Initial installation and hookups shall be made free of charge to all subscribers to encourage maximum system saturation and the earliest possible use of the system for nonentertainment programming purposes.

Construction and Service

1. Construction of each CATV system must be completed within 18 months from the date of franchise award and system connections made available to all potential subscribers in the franchise area. The franchisee shall submit a construction and installation schedule to the commission within 90 days after the franchise award. If this provision is not adopted, poor, minority, and isolated neighborhoods may be denied services indefinitely.

2. All repairs must be made within two days of a customer's complaint.

Future Demands The franchisee is required to install systems in excess of the foreseeable demand for public channels so that at no time is there a danger of a shortage of communications lines.

Performance Review At no longer than three-year intervals, the agency or commission will hold public hearings to review the performance of the franchisee. Within a reasonable time after the conclusion of the hearing, the franchisee must correct deficiencies and improve his services to meet and satisfy all standards and conditions specified by the Commission. Failure to comply shall constitute cause for revocation of the franchise agreement.

Programming Requirements

Local Program Organization Franchise applicants should be required to state their plans for providing locally produced programs to subscribers. Much of the social and economic potential of cable television is related to the capacity of the system to carry a wide range of locally produced nonentertainment programming such as job, health, and education information; civic, social and government meetings, and other socially useful programs.

Public Access At least 50% of all channels should be open for public use on a lease basis. A minimum of one free channel shall be provided for each distinct neighborhood. Accessability to channels shall be open to all regardless of race, sex, national origin, religion, creed, arrest or conviction record.

Program Facilities The franchisee shall provide adequate studio facilities without charge to accommodate persons using the free channel or leasing channel time for either live or video-taped programs.

Technical Assistance The franchisee shall provide free technical assistance to persons using the free channel or leasing channel time.

Scheduling Cablecasts Each franchisee shall provide channel capacity in convenient time segments for cablecasting use by the public in a first-come, first-served common carrier basis.

Control of Program Content The franchisee will not exercise any control over program content nor be liable for that content.

Minority Business Opportunity Requirements

Set-Aside of Franchise Areas The minority community in each of the cities in the Top 100 Markets should receive the opportunity to own and control the cable system in at least one franchise area or cable district. A set-aside provision, similar to that used to promote small business development, should be adopted for this purpose. The minority franchise area should be proportional to all other franchise areas in terms of households and average homes per mile.

Purchase of Supplies, Equipment, and Services Franchise applicants should be required to stipulate in writing their plans to purchase supplies, equipment, and services from local and national minority-owned and operated businesses, including banking, construction, personnel, advertising, janitorial, printing, real estate transactions, computer and similar services available from minority firms. An implementation plan shall be filed with the Commission.

Fair Employment Practices

Minority Employment and Training

1. The franchisee shall recruit and train employees so that minority groups are represented in its employee work force in the same relative proportion as they are represented in the population of the franchise area.

2. Each franchisee shall file an affirmative action plan to this end annually with the Commission. This report shall include a report of persons employed together with their positions and salaries by categories.

3 The franchisee shall not discriminate in hiring or promoting employees on the basis of race, sex, national origin, religion, creed or arrest or conviction records.

Collective Bargaining Each franchisee shall recognize and bargain with any representative body, union or association of employees that the employees by majority vote choose to have as their bargaining agent.

Protection of Privacy Rights

Unauthorized Surveillance and Monitoring It shall be a felony, punishable by jail and/or fine to tap a system line without authorization from the parties whose communication might be overheard.

No monitoring of any terminal connected to the system shall take place without specific written authorization by the user of the terminals in question on each occasion.

In no event shall monitoring of any kind take place without a clearly visible light signal and clearly audible sound signal. The light shall be visible and the sound audible at a distance of at least thirty feet from the terminal at the time of monitoring.

All terminals shall be equipped with a switch by which the user can, upon notification by means of the aforementioned light and sound, prevent the monitoring of his terminal notwithstanding any prior agreement.

fcc rules and regulations governing cable television and new york city rules governing access to public channels

As pointed out in Section One, Chapters I and II, it is anticipated that the FCC will adopt new rules for regulating cable television late this year (1971) and that the new rules will become effective early in 1972. Until new rules are adopted, the rules set forth in this Section must be followed. These rules are contained in Volume III, Part 74, subpart K of the FCC Rules and Regulations. Copies of the entire volume can be obtained for $7.00 from the U.S. Government Printing Office, Washington, D.C. 20402.

The New York City Rules on access to public channels are likely to be used as a model by other city governments in large cities as cable systems expand and public interest groups demand that channels be allotted for cablecasting and "soapbox" use by individuals, groups, agencies and organizations. A detailed discussion on local program origination and cablecasting is contained in Section One, Chapter III.

fcc rules and regulations governing cable television

§ 74.1101 Definitions.

(a) *Community antenna television system.* The term "community antenna television system" ("CATV system") means any facility which, in whole or in part, receives directly or indirectly over the air and amplifies or otherwise modifies the signals transmitting programs broadcast by one or more television stations and distributes such signals by wire or cable to subscribing members of the public who pay for such service, but such term shall not include (1) any such facility which serves fewer than 50 subscribers, or (2) any such facility which serves only the residents of one or more apartment dwellings under common ownership, control, or management, and commercial establishments located on the premises of such an apartment house.

(b) *Television station; television broadcast station; television translator station.* The terms "television station" and "television broadcast station" mean any television broadcasting station operating on a channel regularly assigned to its community by § 73.606 of this chapter. The term "television translator station" means a television broadcast translator station as defined in § 74.701 of this chapter. A television translator station which is licensed to and rebroadcasts the programing of a television broadcast station within that station's Grade B contour, shall be deemed an extension of the originating station.

(c) *Principal community contour.* The term "principal community contour" means the signal contour which a television station is required to place over its entire principal community by § 73.685(a) of this chapter.

(d) *Grade A and Grade B contours.* The terms "Grade A contour" and "Grade B contour" mean the field intensity contours defined in § 73.683(a) of this chapter.

(e) *Network programing.* The term "network programing" means the programing supplied by a national television network organization.

(f) *Substantially duplicated.* The term "substantially duplicated" means regularly duplicated by the network programing of one or more stations, singly or collectively, in a normal week during the hours of 6 to 11 p.m., local time, for a total of 14 or more hours.

(g) *Priority.* The term "priority" means the priority among stations established in § 74.1103(a).

(h) *Independent station.* The term "independent station" means a television station which is not affiliated with any national television network organization.

(i) *Distant signal.* The term "distant signal" means the signal of a television broadcast station which is extended or received beyond the Grade B contour of that station.

(j) *Cablecasting.* The term "cablecasting" means programing distributed on a CATV system which has been originated by the CATV operator or by another entity, exclusive of broadcast signals carried on the system.

(k) *Legally qualified candidate.* The term "legally qualified candidate" means any person who has publicly announced that he is a candidate for nomination by a convention of a political party or for nomination or election in a primary, special, or general election, municipal, county, State, or National, and who meets the qualifications prescribed by the applicable laws to hold the office for which he is a candidate, so that he may be voted for by the electorate directly or by means of delegates or electors, and who:

(1) Has qualified for a place on the ballot, or

(2) Is eligible under the applicable law to be voted for by sticker, by writing his name on the ballot, or other method, and (i) has been duly nominated by a political party which is commonly known and regarded as such, or (ii) makes a substantial showing that he is a bona fide candidate for nomination or office.

[*§ 74.1101 pars (j) & (k) adopted eff. 12-1-69; III (68)-6*]

§ 74.1103 Requirements relating to distribution of television signals by community antenna television systems.

No community antenna television system shall supply to its subscribers signals broadcast by one or more television stations, except in accordance with the following conditions:

(a) *Stations required to be carried.* Within the limits of its channel capacity, any such CATV system shall carry the signals of operating or subsequently authorized and operating television broadcast and 100 watts or higher power translator stations in the following order of priority, upon the request of the licensee or permittee of the relevant station:

(1) First, all commercial and noncommercial educational stations within whose principal community contours the system or the community of the system is located, in whole or in part;

(2) Second, all commercial and noncommercial educational stations within whose Grade A contours the system or the community of the system is located, in whole or in part;

(3) Third, all commercial and noncommercial educational stations within whose Grade B contours the system or the community of the system is located, in whole or in part; and

(4) Fourth, all commercial and noncommercial educational translator stations operating in the community of the system, in whole or in part, with 100 watts or higher power.

(b) *Exceptions.* Nothwithstanding the requirements of paragraph (a) of this section,

(1) The system need not carry the signal of any station, if (i) that station's network programing is substantially duplicated by one or more stations of higher priority, and (ii) carrying it would, because of limited channel capacity, prevent the system from carrying the signal of an independent commercial station or a noncommercial educational station.

(2) In cases where (i) there are two or more signals of equal priority which substantially duplicate each other, and (ii) carrying all such signals would, because of limited channel capacity, prevent the system from carrying the signal of an independent commercial station or a noncommercial educational station, the system need not carry all such substantially duplicat-

ing signals, but may select among them to the extent necessary to preserve its ability to carry the signals of independent commercial or noncommercial educational stations.

(3) The system need not carry the signal of any television translator station if: (i) The system is carrying the signal of the originating station, or (ii) the system is within the Grade B or higher priority contour of a station carried on the system whose programing is substantially duplicated by the translator; *Provided, however*, That where the originating station is carried in place of the translator station, the priority for purposes of paragraph (e) of this section shall be that of the translator station unless the priority of the originating station is higher.

(4) In the event that the system operates, or its community is located, within the Grade B or higher priority contours of both a satellite and its parent station, the system need carry only the station with the higher priority, if the satellite station and its parent station are of equal priority, the system may select between them.

(c) *Special requirements in the event of noncarriage.* Where the system does not carry the signals of one or more stations within whose Grade B or higher priority contour it operates, or the signals of one or more 100 watts or higher power translator stations located in its community, the system shall offer and maintain, for each subscriber, an adequate switching device to allow the subscriber to choose between cable and noncable reception, unless the subscriber affirmatively indicates in writing that he does not desire this device.

(d) *Manner of carriage.* Where the signal of any station is required to be carried under this section,

(1) The signal shall be carried without material degradation in quality (within the limitations imposed by the technical state of the art);

(2) The signal shall, upon request of the station licensee or permittee, be carried on the system on the channel on which the station is transmitting (where practicable without material degradation); and

(3) The signal shall, upon the request of the station licensee or permittee, be carried on the system on no more than one channel.

(e) *Stations entitled to program exclusivity.* Any such system which operates, in whole or in part, within the Grade B or higher priority contour of any commercial or noncommercial educational television station or within the community of a fourth priority television translator station, and which carries the signal of such station shall, upon request of the station licensee or permittee, maintain the station's exclusivity as a program outlet against lower priority or more distant duplicating signals, but not against signals of equal priority, in the manner and to the extent specified in paragraphs (f) and (g) of this section.

(f) *Program exclusivity; extent of protection.* Where a station is entitled to program exclusivity, the CATV system shall, upon the request of the station licensee or permittee, refrain from duplicating any program broadcast by such station, on the same day as its broadcast by the station, if the CATV operator has received notification from the requesting station of the date and time of its broadcast of the program and the date and time of any broadcast to be deleted, as soon as possible and in any event no later than 48 hours prior to the broadcast to be deleted. Upon request of the CATV system, such notice shall be given at least 8 days prior to the date of any broadcast to be deleted.

(g) *Exceptions.* Notwithstanding the requirements of paragraph (f) of this section.

(1) The CATV system need not delete reception of a network program if, in so doing, it would leave available for reception by subscribers, at any time, less than the programs of two networks (including those broadcast by any stations whose signals are being carried and whose program exclusivity is being protected pursuant to the requirements of this section);

(2) The system need not delete reception of a network program which is scheduled by the network between the hours of 6 and 11 p.m., eastern time, but is broadcast by the station requesting deletion, in whole or in part, outside of the period which would normally be considered prime time for network programing in the time zone involved;

(3) The system need not delete reception of any program consisting of the broadcast coverage of a speech or other event as to which the time of presentation is of special significance, except where the program is being simultaneously broadcast by a station entitled to program exclusivity; and

(4) The system need not delete reception of any program which would be carried on the system in color but will be broadcast in black and white by the station requesting deletion.

§ 74.1105 Notification prior to the commencement of new service.

(a) No CATV system shall commence operations in a community or commence supplying to its subscribers the signal of any television broadcast station carried beyond the Grade B contour of the station, unless the system has given prior notice of the proposed new service to the licensee or permittee of any television broadcast station within whose predicted Grade B contour the system operates or will operate, and to the licensee or permittee of any 100 watts or higher power translator station operating in the community of the system, and has furnished a copy of each such notification to the Federal Communications Commission, within sixty (60) days after obtaining a franchise or entering into a lease or other arrangement to use facilities; in any event, no CATV system shall commence such operations until thirty (30) days after notice has been given. Such notice shall be given by existing systems which propose to add new distant signals at least thirty (30) days prior to commencing service and by systems which propose to extend lines into another community within sixty (60) days after obtaining a franchise or entering into a lease or other arrangement to use facilities or where no new local authorization or contractual arrangement is necessary, at least thirty (30) days prior to commencing service. Where it is proposed to extend the signal of any noncommercial educational television station beyond its Grade B contour into a community with an unoccupied reserved educational television channel assignment under § 73.606 of this chapter, the notice shall also be served upon the superintendents of schools in the community and county in which the system will operate and the local, area, and State educational television agencies, if any.

(b) The notice shall include the name and address of the system, identification of the community to be served, the television signals to be distributed, and the estimated time operations will commence.

(c) Where a petition with respect to the proposed service is filed with the Commission, pursuant to § 74.1109 of this chapter, within thirty (30) days after notice, new service which is challenged in the petition shall not be commenced until after the Commission's ruling on the petition or on the interlocutory question of temporary relief pending further procedures; *Provided, however,* That service shall not be commenced in violation of the terms of any specified temporary relief or of the provisions of § 74.1107 of this chapter. Where no petition pursuant to § 74.1109 has been filed within thirty (30) days after notice, service may be commenced at any time thereafter, subject, however, to the provisions of § 74.1107.

(d) The provisions of this section do not apply to any signals which were being supplied to subscribers in the community of the CATV system on March 17, 1966, unless it is proposed to extend lines into another community.

NOTE 1: As used in § 74.1105, the term "predicted Grade B contour" means the field intensity contour defined in § 73.683(a) of this chapter, the location of which is determined exclusively by means of the calculations prescribed in § 73.684 of this chapter.

NOTE 2: As used in § 74.1105, the term "television broadcast station" includes foreign television broadcast stations.

§ 74.1107 Requirement for showing in evidentiary hearing and Commission approval in top 100 television markets; other procedures.

(a) No CATV system operating in a community within the predicted Grade A contour of a television broadcast station in the 100 largest television markets shall extend the signal of a television broadcast station beyond the Grade B contour of that station, except upon a showing approved by the Commission that such extension would be consistent with the public interest, and specifically the establishment and healthy maintenance of television broadcast service in the area. Commission approval of a request to extend a signal in the foregoing circumstances will be granted where the Commission, after consideration of the request and all related materials in a full evidentiary hearing, determines that the requisite showing has been made. The market size shall be determined by the rating of the American Research Bureau, on the basis of the net weekly circulation for the most recent year.

(b) A request under paragraph (a) of this section shall be filed after the CATV system has obtained any necessary franchise for operation or has entered into a lease or other arrangement to use facilities and shall set forth the name of the community involved, the date on which a franchise was obtained, the signal or signals proposed to be extended beyond their Grade B contours, the date on which copies of the notifications required by § 74.1105 of this chapter were filed with the Commission, and the specific reasons why it is urged that such extension is consistent with the public interest. Public notice will be given of the filing of such a request, and interested parties may file a response or statement within thirty (30) days after such public notice. A reply to such a response or statement may be filed within a twenty (20) day period thereafter. The Commission shall designate the request for an evidentiary hearing on issues to be specified, with the burden of proof and the burden of proceeding with the introduction of evidence upon the CATV system making the request, unless otherwise specified by the Commission as to particular issues.

(c) No CATV system, located so as to fall outside the provisions of paragraph (a) of this section, shall extend the signal of a television broadcast station beyond the Grade B contour of that station, where the Commission, upon its own motion or pursuant to a petition filed under § 74.1109, determines, after appropriate proceedings, that such extension would be inconsistent with the public interest, taking into account particularly the establishment and healthy maintenance of television broadcast service in the area.

(d) The provisions of paragraphs (a) and (b) of this section shall not be applicable to any signals which were being supplied by a CATV system to its subscribers in a community on February 15, 1966, and pursuant to a franchise (where necessary) issued on or before that date: *Provided, however,* That any new franchise or amendment of an existing franchise after February 15, 1966, to operate or extend the operations of the CATV system in the same general area or any extension into another community does come within the provisions of paragraphs (a) and (b) of this section: *And provided further,* That no CATV system located in a community in the 100 largest television markets, which was supplying to its subscribers on February 15, 1966, a signal carried beyond its Grade B contour, shall extend such service to new geographical areas within the same community where the Commission, upon petition filed under § 74.1109 by a television broadcast station or other interested person located in the area and after consideration of the response of the CATV system and appropriate proceedings, determines that the public interest, taking into account the considerations set forth in the Second Report and Order in Docket Nos. 14895, 15233, and 15971, FCC 66–220, paragraphs 113–149, would be served by appropriate conditions limiting the geographical extension of the system to new areas in the community. The Commission may also consider, upon the basis of the pleadings before it, whether temporary relief is called for in the public interest, and, if so, the nature of such relief; no CATV system coming within the foregoing provision shall extend its service to new geographical areas in violation of the terms of the specified temporary relief.

(e) Within 60 days of issuance of a request filed pursuant to paragraph (a) of this section, interested parties seeking simultaneous consideration with such request must file appropriate requests for any other CATV system in the same television market. All requests for CATV systems in a given market timely filed with respect to the first request will be processed and considered simultaneously. Later filed requests for the particular market will be subject to chronological processing and may not be considered in the same proceeding as the earlier requests.

NOTE 1: As used in § 74.1107, the term "television broadcast station" includes foreign television broadcast stations.

§ 74.1109 Procedures applicable to petitions for waiver of the rules, additional or different requirements and rulings on complaints or disputes.

(a) Upon petition by a CATV system, an applicant, permittee, or licensee of a television broadcast, translator, or microwave relay station, or by any other interested person, the Commission may waive any provision of the rules relating to the distribution of television broadcast signals by CATV systems, impose additional or different requirements, or issue a ruling on a complaint or disputed question.

(b) The petition may be submitted informally, by letter, but shall be accompanied by an affidavit of service on any CATV system, station licensee, permittee, applicant, or other interested person who may be directly affected if the relief requested in the petition should be granted.

(c) (1) The petition shall state the relief requested and may contain alternative requests. It shall state fully and precisely all pertinent facts and considerations relied upon to demonstrate the need for the relief requested and to support a determination that a grant of such relief would serve the public interest. Factual allegations shall be supported by affidavit of a person or persons with actual knowledge of the facts, and exhibits shall be verified by the person who prepares them.

(2) A petition for a ruling on a complaint or disputed question shall set forth all steps taken by the parties to resolve the problem, except where the only relief sought is a clarification or interpretation of the rules.

(d) Interested persons may submit comments or opposition to the petition within thirty (30) days after it has been filed. Upon good cause shown in the petition, the Commission may, by letter or telegram to known interested persons, specify a shorter time for such submissions. Comments or oppositions shall be served on petitioner and on all persons listed in petitioner's affidavit of service, and shall contain a detailed full showing, supported by affidavit, of any facts or considerations relied upon.

(e) The petitioner may file a reply to the comments or oppositions within twenty (20) days after their submission, which shall be served upon all persons who have filed pleadings and shall also contain a detailed full showing, supported by affidavit, of any additional facts or considerations relied upon. Upon good cause shown, the Commission may specify a shorter time for the filing of reply comments.

(f) The Commission, after consideration of the pleadings, may determine whether the public interest would be served by the grant, in whole or in part, or denial of the request, or may issue a ruling on the complaint or dispute. The Commission may specify other procedures, such as oral argument, evidentiary hearing, or further written submissions directed to particular aspects, as it deems appropriate. In the event that an evidentiary hearing is required, the Commission will determine, on the basis of the pleadings and such other procedures as it may specify, whether temporary relief should be accorded to any party pending the hearing and the nature of any such temporary relief. Where a petition involves new service to subscribers (other than service coming within the provisions of § 74.1107(a) of this chapter), the Commission will expedite its consideration and promptly issue a ruling either on the merits of the petition or on the interlocutory question of temporary relief pending further procedures.

(g) Where a request for temporary relief is contained in a petition with respect to service coming within the provisions of § 74.1107(d) of this chapter, opposition to such request for temporary relief shall be filed within ten (10) days and reply comments within seven (7) days thereafter. The commission will expedite its consideration of the question of temporary relief.

(h) Where a petition for waiver of the provisions of § 74.1103(a) of this chapter is filed within fifteen (15) days after a request for carriage, the system need their authorized spokesmen, or those associated with them in the campaign, on other such candidates, their authorized spokesmen, or persons associated with the candidates in the campaign; and (3) to bona fide newscasts, bona fide news interviews, and on-the-spot coverage of a bona fide news event (including commentary or analysis contained in the foregoing programs, but the provisions of paragraph (b) of this section shall be applicable to editorials of the licensee).

(d) Where a CATV system, in an editorial, (1) endorses or (2) opposes a legally qualified candidate or candidates, the system shall, within 24 hours after the editorial, transmit to respectively (i) the other qualified candidate or candidates for the same office or (ii) the candidate opposed in the editorial (*a*) notification of the date, time, and channel of the editorial; (*b*) a script or tape of the editorial; and (*c*) an offer of a reasonable opportunity for a candidate or a spokesman of the candidate to respond over the system's facilities; *Provided, however,* That where such editorials are cablecast within 72 hours prior to the day of the election, the systems shall comply with the provisions of this paragraph sufficiently far in advance of the broadcast to enable the candidate or candidates to have a reasonable opportunity to prepare a response and to present it in a timely fashion.

§ 74.1116 Lotteries.

(a) No CATV system when engaged in cablecasting shall transmit or permit to be transmitted on the cablecasting channel or channels any advertisement of or information concerning any lottery, gift enterprise, or similar scheme, offering prizes dependent in whole or in part upon lot or chance, or any list of the prizes drawn or awarded by means of any such lottery, gift enterprise, or scheme, whether said list contains any part or all of such prizes.

(b) The determination whether a particular program comes within the provisions of paragraph (a) of this section depends on the facts of each case. However, the Commission will in any event consider that a program comes within the provisions of paragraph (a) of this section if in connection with such program a prize consisting of money or thing of value is awarded to any person whose selection is dependent in whole or in part upon lot or chance, if as a condition of winning or competing for such prize, such winner or winners are required to furnish any money or thing of value or are required to have in their possession any product sold, manufactured, furnished or distributed by a sponsor of a program broadcast on the station in question.

[§ *74.1116 added eff. 8–14–70; III (68)–9*]

§ 74.1117 Advertising.

A CATV system engaged in cablecasting may present advertising material at the beginning and conclusion of each cablecast program and at natural intermissions or breaks within a cablecast, *Provided*, That the system itself does not interrupt the presentation of program material in order to intersperse advertising: *And provided, further*, That advertising material is not presented on or in connection with cablecasting in any other manner.

NOTE: The term "natural intermissions or breaks within a cablecast" means any natural intermission in the program material which is beyond the control of the CATV operator, such as time-out in a sporting event, an intermission in a concert or dramatic performance, a recess in a city council meeting, an intermission in a long motion picture which was present at the time of theatre exhibition, etc.

§ 74.1119 Sponsorship identification.

(a) When a CATV system engaged in cablecasting presents any matter for which money, services, or other valuable consideration is either directly or indirectly paid or promised to, or charged or received by, such system, the system shall make an announcement that such matter is sponsored, paid for, or furnished, either in whole or in part, and by whom or on whose behalf such consideration was supplied; *Provided, however*, That "service or other valuable consideration" shall not include any service or property furnished without charge or at a nominal charge for use on, or in connection with, a cablecast unless it is so furnished for consideration for an identification in a cablecast of any person, product, service, trademark, or brand name beyond an identification which is reasonably related to the use of such service or property on the cablecast.

(b) Each system engaged in cablecasting shall exercise reasonable diligence to obtain from its employees, and from other persons with whom it deals directly in connection with any program matter for cablecasting, information to enable it to make the announcement required by this section.

(c) In the case of any political program or any program involving the discussion of public controversial issues for which any films, records, transcriptions, talent, script, or other material or services of any kind are furnished, either directly or indirectly, to a CATV system as an inducement to the cablecasting of such program, an announcement to this effect shall be made at the beginning and conclusion of such program: *Provided, however*, That only one such announcement need be made in the case of any such program of 5 minutes' duration or less, either at the beginning or conclusion of the program.

(d) The announcements required by this section are waived with respect to feature motion picture films produced initially and primarily for theatre exhibition.

§ 74.1121 General operating requirements.

(a) Cablecasting operations for which a per-program or per-channel charge is made shall comply with the following requirements:

(1) Feature films shall not be cablecast which have had general release in theaters anywhere in the United States more than 2 years prior to their cablecast: *Provided, however*, That during 1 week of each calendar month one feature film the general release of not carry the signal of the requesting station pending the Commission's ruling on the petition or on the interlocutory question of temporary relief pending further procedures.

§ 74.1111 Cablecasting in conjunction with carriage of broadcast signals.

(a) Effective on and after April 1, 1971, no CATV system having 3,500 or more subscribers shall carry the signal of any television broadcast station unless the system also operates to a significant extent as a local outlet by cablecasting and has available facilities for local production and presentation of programs other than automated services: *Provided, further*, That the system shall not enter into any contract, arrangement or lease for use of its cablecasting facilities which prevents or inhibits the use of such facilities for a substantial portion of time (including the time period 6–11 p.m.), for local programing designed to inform the public on controversial issues of public importance.

(b) No CATV system shall carry the signal of any television broadcast station if the system engages in cablecasting, either voluntarily or pursuant to paragraph (a) of this section, unless such cablecasting is conducted in accordance with the provisions of §§ 74.1113, 74.1115, 74.1117, and 74.1119.

[§ *74.1111(a) revised eff. 8–14–70; III(68)–9*]

§ 74.1113 Cablecasts by candidates for public office.

(a) *General requirements*. If a CATV system shall permit any legally qualified candidate for public office to use its cablecasting facilities, it shall afford equal opportunities to all other such candidates for that office to use such facilities: *Provided*, That such system shall have no power of censorship over the material cablecast by any such candidate: *And provided further*, That an appearance by a legally qualified candidate on any:

(1) Bona fide newscast,

(2) Bona fide news interview,

(3) Bona fide news documentary (if the appearance of the candidate is incidental to the presentation of the subject or subjects covered by the news documentary), or

(4) On-the-spot coverage of bona fide news events (including but not limited to political conventions and activities incidental thereto),

shall not be deemed to be use of the facilities of the system within the meaning of this paragraph.

NOTE: The fairness doctrine is applicable to these exempt categories. See § 74.1115.

(b) *Rates and practices*. (1) The rates, if any, charged all such candidates for the same office shall be uniform, shall not be rebated by any means direct or indirect, and shall not exceed the charges made for comparable use of such facilities for other purposes.

(2) In making facilities available to candidates for public office no CATV system shall make any discrimination between candidates in charges, practices, regulations, facilities, or services for or in connection with the service rendered, or make or give any preference to any candidate for public office or subject any such candidate to any prejudice or disadvantage; nor shall any CATV system make any contract or other agreement

which shall have the effect of permitting any legally qualified candidate for any public office to cablecast to the exclusion of other legally qualified candidates for the same public office.

(c) *Records, inspections.* Every CATV system shall keep and permit public inspection of a complete record of all requests for cablecasting time made by or on behalf of candidates for public office, together with an appropriate notation showing the disposition made by the system of such requests, and the charges made, if any, if the request is granted. Such records shall be retained for a period of 2 years.

(d) *Time of request.* A request for equal opportunities must be submitted to the CATV system within 1 week of the day on which the first prior use, giving rise to the right to equal opportunities, occurred: *Provided, however,* That where a person was not a candidate at the time of such first prior use, he shall submit his request within 1 week of the first subsequent use after he has become a legally qualified candidate for the office in question.

(e) *Burden of proof.* A candidate requesting such equal opportunities of the CATV system, or complaining of noncompliance to the Commission, shall have the burden of proving that he and his opponent are legally qualified candidates for the same public office.

[§ *74.1113(d) revised eff. 6-8-70; III(68)-9***]**

§ 74.1115 Fairness Doctrine; personnal attacks; political editorials.

(a) A CATV system engaging in cablecasting shall afford reasonable opportunity for the discussion of conflicting views on issues of public importance.

NOTE: See public notice: Applicability of the Fairness Doctrine in the Handling of Controversial Issues of Public Importance, 29 F.R. 10415.

(b) When, during the presentation of views on a controversial issue of public importance, an attack is made upon the honesty, character, integrity, or like personal qualities of an identified person or group, the CATV system shall, within a reasonable time and in no event later than 1 week after the attack, transmit to the person or group attacked (1) notification of the date, time, and identification of the cablecast; (2) a script or tape (or an accurate summary if a script or tape is not available) of the attack; and (3) an offer of a reasonable opportunity to respond over the system's facilities.

(c) The provisions of paragraph (b) of this section shall not be applicable (1) to attacks on foreign groups or foreign public figures; (2) to personal attacks which are made by legally qualified candidates, which occurred more than 10 years previously may be cablecast, and more than a single showing of such film may be made during that week: *Provided, further,* That feature films the general release of which occurred between 2 and 10 years before proposed cablecast may be cablecast upon a convincing showing to the Commission that bona fide attempt has been made to sell the films for conventional television broadcasting and that they have been refused, or that the owner of the broadcast rights to the films will not permit them to be televised on conventional television because he has been unable to work out satisfactory arrangements concerning editing for presentation thereon, or perhaps because he intends never to show them on conventional television since to do so might impair their repetitive box office potential in the future.

NOTE: As used in this subparagraph, "general release" means the first-run showing of a feature film in a theater or theaters in an area, on a nonreserved-seat basis, with continuous performances. For first-run showing of feature films on a nonreserved-seat basis which are not considered to be 'general release" for purposes of this subparagraph, see note 56 in the fourth report and order in Docket No. 11279, 15 FCC 2d 466.

(2) Sports events shall not be cablecast which have been televised live on a nonsubscription, regular basis in the community during the 2 years preceding their proposed cablecast: *Provided, however,* That if the last regular occurrence of a specific event (e.g., summer Olympic games) was more than 2 years before proposed showing on CATV in a community and the event was at that time televised on conventional television in that community, it shall not be cablecast.

NOTE 1: In determining whether a sports event has been televised in a community on a nonsubscription basis, only commercial television broadcast stations which place a Grade A contour over the entire community will be considered. Such stations need not necessarily be licensed to serve that community.

NOTE 2: The manner in which this subparagraph will be administered and in which "sports," "sports events," and "televised live on a nonsubscription regular basis" will be construed is explained in paragraphs 288–305 of the fourth report and order in Docket No. 11279, 15 FCC 2d 466.

(3) No series type of program with interconnected plot or substantially the same cast of principal characters shall be cablecast.

(4) Not more than 90 percent of the total cablecast programing hours shall consist of feature films and sports events combined. The percentage calculations may be made on a yearly basis, but, absent a showing of good cause, the percentage of such programing hours may not exceed 95 percent of the total cablecast programing hours in any calendar month.

(5) No commercial advertising announcements shall be carried on such channels during such operations except, before and after such programs, for promotion of other programs for which a per-program or per-channel charge is made.

[§ *74.1121 added eff. 8-14-70; III(68)-9***]**

§ 74.1131 Diversification of control over communications media.

(a) *Cross-ownership.* No CATV system (including all parties under common control) shall carry the signal of any television broadcast station if such system directly or indirectly owns, operates, controls, or has an interest in:

(1) A national television network (such as ABC, CBS, or NBC); or

(2) A television broadcast station whose predicted Grade B contour, computed in accordance with § 73.684 of this chapter, overlaps in whole or in part the service area of such system (i.e., the area within which the system is serving subscribers); or

(3) A television translator station authorized to

serve a community within which the system is serving subscribers.

NOTE 1: The word "control" as used herein is not limited to majority stock ownership, but includes actual working control in whatever manner exercised.

NOTE 2: The word "interest' as used herein includes, in the case of corporations, common officers or directors and partial (as well as total) ownership interests represented by ownership of voting stock.

NOTE 3: In applying the provisions of paragraph (a) of this section to the stockholders of a corporation which has more than 50 stockholders:

(a) Only those stockholders need be considered who are officers or directors or who directly or indirectly own 1 percent or more of the outstanding voting stock.

(b) Stock ownership by an investment company, as defined in 15 U.S.C. section 80a–3 (commonly called a mutual fund), need be considered only if it directly or indirectly owns 3 percent or more of the outstanding voting stock or if officers or directors of the corporation are representatives of the investment company. Holdings by investment companies under common management shall be aggregated. If an investment company directly or indirectly owns voting stock in an intermediate company which in turn directly or indirectly owns 50 percent or more of the voting stock of the corporation, the investment company shall be considered to own the same percentage of outstanding shares of such corporation as it owns of the intermediate company: *Provided, however,* That the holding of the investment company need not be considered where the intermediate company owns less than 50 percent of the voting stock, but officers or directors of the corporation who are representatives of the intermediate company shall be deemed to be representatives of the investment company.

(c) In cases where record and beneficial ownership of voting stock is not identical (e.g., bank nominees holding stock as record owners for the benefit of mutual funds, brokerage houses holding stock in street name for the benefit of customers, trusts holding stock as record owners for the benefit of designated parties), the party having the right to determine how the stock will be voted will be considered to own it for the purposes of this section.

(d) Effective date: The provisions of paragraph (a) of this section are not effective until August 10, 1973, as to ownership interests proscribed herein if such interests were in existence on or before July 1, 1970 (e.g., if a franchise were in existence on or before July 1, 1970): *Provided, however,* That the provisions of paragraph (a) of this section are effective on August 10, 1970, as to such interests acquired after July 1, 1970.

[§ *74.1131 added eff. 8–10–70; III(68)–9*]

---

new york city rules governing access to public channels

The following rules shall apply to the availability of, and access to, Public Channels for the interim period of July 1–December 31, 1971. The rules may be amended without notice by the Director of Franchise. The rules are intended to provide guidelines, and are not expected to cover every contingency that may arise. It is anticipated that the rules will be revised when the City, the CATV companies and the public have some meaningful experence with the Public Channels.

For the purpose of gaining such experience and in order to encourage differing uses of the Public Channels the two Public Channels shall be governed by different concepts. On one Public Channel, denominated Channel C in the franchise, there shall be an opportunity to reserve a particular time period each week, or several time periods each week, in order to permit the user to build an audience on a regular basis. On the other Public Channel, denominated Channel D in the franchise, there shall be no multiple time reservations, in order to permit a user with a single program and users with relatively last-minute requirements access to prime time periods.

During the interim period, users are invited to suggest changes in these rules. All suggested rule changes, with the reasons therefor, are to be mailed to the Director of Franchises, 1307 Municipal Building, New York, New York 10007.

The following shall apply to all Public Channels:

1. The user shall permit the Company to preview the programs it wishes to present on a Public Channel so that a determination may be made as to whether the cablecasting of any such program will subject the Company to liability under applicable law. For the interim period covered by these rules, that determination shall be made by the Company. Should the user dispute the Company's determination, both the user and the Company shall submit to the Director of Franchises a written statement setting forth the facts surrounding the dispute and their views on the matter. Such statements shall be for informational purposes only and the Director shall not interfer with the Company's determination. The Company's determination shall, of course, be subject to judicial review if the user chooses to assert such a legal challenge. If the Company determines that any program, or part thereof, is objectionable in that it will subject it to liability, the user shall be given an opportunity to revise the program so as to delete the objectionable portion. If the user chooses not to do so, he may withdraw the program. Otherwise, the Company shall transmit the program after itself deleting any objectionable portion thereof unless in its judgment the entire program is objectionable. If the entire program is objectionable or is withdrawn, any fees paid for channel time shall be returned to the user.

2. The user shall submit to the Company an application for channel time at least two weeks in advance of the requested time period. Nevertheless, if an applicant with programming as to which timeliness is a factor requests time on Channel D on less than two weeks' notice, the Company shall devote its best efforts to clearing the program pursuant to Rule 1 in order to comply with the user's request. In addition, for the month of July, 1971, the Company shall make every effort to accommodate users of both channels on shorter notice.

3. Every application shall contain the following information:

(a) Requested date and hour of use:;

(b) Length of program:

(c) A general indication of the purpose of the program and a statement as to whether any commercial material will be included in it;

(d) A list of individuals who will appear on the program;

(e) Name, address and telephone number of the individual or organization making the reuqest, and if an organization, the names of the principal officers;

(f) The method by which the program is to be presented;

(g) Whether studio and/or production facilities are requested.

4. No minor under 18 years of age shall be permitted use of the Company's facilities unless accompanied by an adult, who shall assume all legal responsibility for the program and the actions of the minors and shall be responsible for obtaining whatever permits may be required authorizing the appearance of such minors.

5. At least ten days prior to the date on which the program is to be carried on a public channel the applicant must:

(a) Read, execute and file with the company its public channel contract, in which the company may specify reasonable technical standards;

(b) Where music is included, furnish to the company the title of the music, if any; the name of the composer(s); the licensing agent for performance rights; and appropriate documents authorizing performance on the public channel;

(c) Where non-music copyrighted material is included in any program, furnish the company the name of the author, the copyright owner and appropriate documents authorizing the use of the material on the program.

6. Applications for time on Channel C shall be granted on a first-come, first-served basis, subject to the following qualifications:

(a) A user may lease no more than two hours (cumulative) per week of prime time (7:00-11:00 P.M.).

(b) A user may lease no more than 7 hours per week of prime and non-prime time, unless there are no other requests for the time periods beyond such 7 hours.

7. Applications for time on Channel D shall be granted on a first-come, first-served basis, subject to the following qualifications:

(a) A user may not make advance reservations of the same time period more than once per month.

(b) Up to one week prior to the time period requested, users who have leased 5 hours or more on both Public Channels within the previous month shall be subject to scheduling displacement by less frequent users, regardless of the timing of their applications.

8. No brokering will be permitted nor will agency fees or commissions of any kind be payable by the Company unless expressly agreed to in writing.

9. Time allocations shall be non-assignable.

10. The rule of decision in all schedule conflicts will be that which provides the greatest diversity of expression.

11. The Company may require users of Public Channels to provide it with all information needed to enable it to comply with applicable rules and regulations of the Federal Communications Commission, or the Company may require users themselves to comply with the aforesaid section.

12. Users who choose to produce their own programming (rather than lease facilities made available by the Company) shall consult with the Company to determine whether they will have to supply any equipment to enable the Company to transmit their programming.

13. Upon request, the Company shall inform users of any production facilities and programming organizations known to the Company which might be of financial and technical assistance to the user.

14. If any user is of the opinion that the Company has treated it unfairly or that the Company has not complied with the aforesaid rules, it should communicate directly with the Director of Franchises.

15. Failure to comply with the aforesaid rules shall subject the user to cancellation of all future reservations of Public Channel time for a two month period.

16. The Company shall make available to all users who request technical assistance the advisory services of a qualified person experienced in television production. Such advisory services shall be provided at the Company's premises and during reasonable business hours.

Issued June 29, 1971 by

Morris Tarshis
Director of Franchises
1307 Municipal Building
New York, N. Y. 10007

NOTE:

Rates for Public Channels are filed by the Companies. They vary according to the particular company's policies.

important federal agencies, private organizations and associations

In addition to the Federal Communications Commission, several federal agencies, congressional committees and private organizations are involved in developing public policies to regulate the cable industry. The principal public and private organizations are listed here. Information about their activities can be obtained from them directly. Trade journals, listed elsewhere in this section also report on the proceedings, interests, and policy positions of these groups.

federal agencies

NAME: Federal Communications Commission

LOCATION: 1919 M Street, N.W.
Washington, D.C. 20554

TELEPHONE: (202) 655-4000

STAFF:		Term Expires
	Chairman: Dean Burch (R-Ariz.)	1976
	Commissioners:	
	Robert T. Bartley (D-Tex.)	1972
	Robert E. Lee (R-Ill.)	1974
	H. Rex Lee (D-D.C.)	1975
	Robert Wells (R-Kan.)	1977
	Nicholas Johnson (D-Iowa)	1973
	Charlotte T. Reid (R-Ill.)	1978

Executive Director: Max D. Paglin
General Counsel: Henry Geller
Secretary: Ben F. Waple
Chief Hearing Examiner: Arthur A. Gladstone

CABLE TELEVISION BUREAU

Room 430
Telephone: (202) 632-6480

Chief: Sol Schildhause
Deputy Chief: Allen Cordon
Legal Staff: Albert J. Baxter
Edward J. Brown
Richard L. Brown
David Goldman
Jerold Jacobs
William H. Johnson
Robert M. Koteen
Abraham A. Leib
Jacob W. Mayer
Walter Morse
Frances V. Peck
John Pellegrin
Stephen Ross
James Tansey
Robert Ungar
Staff Engineer: Early Monroe

The Federal Communications Commission is the United States Government agency charged with regulating interstate and foreign communications by means of radio, television, wire, cable and satellite. The Communications Act of 1934, created the Federal Communications Commission, to be administered by seven commissioners, appointed by the President with the approval of the Senate. The major activities of the Commission are the allocation of bands of frequencies to non-government communications services and assigning frequencies to individual stations; licensing and regulating stations and operators; promoting safety through the use of radio on land, water and in the air; regulating interstate and foreign communications by telegraph, telephone and satellite makes the FCC the principal regulatory body at this point.

In January 1970, because of the rapid growth of cable television, the Commission created a special Cable Television Bureau responsible for administering and enforcing CATV rules.

Rule Making

The FCC develops rules internally when circumstances make some form of regulation necessary on matters under its jurisdiction.

The sequence of procedural actions is as follows:

(a) *"Notice of Proposed Rule Making"* and *"Notice of Inquiry"*. A notice to the general public specifying the intent and area in which rules are being formulated. The public is invited to file comments on the issues raised by the commission.

"Docket numbers" are assigned to the public notice for identification purposes.

(b) *Open Hearings* May be held to permit interested parties to present their position and recommendations concerning the rules.

(c) *Petition for Rulemaking* Rules may also be developed externally by what is called a "petition for rulemaking" where interested parties petition the FCC for rules they feel are necessary, hearings are held, pros and cons are heard and the Commission takes them under advisement and makes a decision to accept the rules or reject them.

(d) *Report and Order* When the Commission has decided which rules will be binding, it issues what is called a "Report and Order," representing the final action taken by the Commission and the rules which are now adopted.

Copies of all rules and regulations are available *free* in the Public Reference Room, (202) 632-7566. Station files for all AM-FM radio, television and CATV licenses are also available for public viewing. This includes sales and renewal records—xerox copies can be made.

NAME: Office of Telecommunication Policy
Executive Office of the President

LOCATION: 1800 G St., N.W.
Washington, D.C. 20504

TELEPHONE: (202) 395-5800

STAFF: Clay T. Whitehead, Director
George F. Mansur, Deputy Director
Wilfrid Dean, Jr., Assistant Director
Walter R. Hinchman, Assistant Director
Charles C. Joyce, Jr. Assistant Director
Antonin Scalia, General Counsel

The Office of Telecommunication Policy was established September 1970 to oversee all government communications, make recommendations and proposals to the President and to serve as the executive branch spokesman for matters in telecommunications.

NAME: White House Cable Television Committee

DESCRIPTION: The President has established this committee to develop forward-looking policy proposals that will permit the full potential of cable TV to be realized and enhance television service.

STAFF: **Chairman:** Clay T. Whitehead, Director of the Office of Telecommunication Policy

Members are:

Elliot L. Richardson, Secretary of HEW
George Romney, Secretary of HUD
Maurice H. Stans, Secretary of Commerce
Robert H. Finch, Counsellor to the President
Leonard Garment, Special Consultant to the President
Herbert G. Klein, Director of Communications for the Executive Branch

Interagency CATV Working Group

Robert Finch's Office
Dr. George Grassmuch
West Wing
The White House
Phone: 395-2888

Health, Education and Welfare
Mr. Albert Horley
Office of Telecommunications
Health, Education & Welfare
Washington, D.C. 20201
Phone: 963-5941

Leonard Garment's Office
Mr. Bradley Patterson
Old Executive Office Bldg.
Room 188½
Phone: 395-2775

Herb Klein's Office
Mr. Alvin Snyder
Assistant to the Director
of Communications
Old Exec. Office Bldg. Rm. 153
Phone: 395-2682

Department of Commerce
Mr. Robert Powers
Office of Telecommunications
Department of Commerce
Washington, D.C. 20230
Phone: 967-3603

Housing and Urban Development
Mr. Alan Siegel
Director of Environmental
Factors and Public Utilities
Dept. of Housing & Urban
Development-Room 4112
451 Seventh Street, S.W.
Washington, D.c. 20410
Phone: 755-5360

Office of Telecommunications Policy
Mr. Walter Hinchman
Assistant Director; 395-5190
Mr. Bruce Owen
Senior Economist; 395-3782

NAME: Office of Economic Opportunity
Communication Development Division M/821

LOCATION: 1200 19th St., N.W.
Washington, D.C. 10506

TELEPHONE: (202) 254-5174

STAFF: Director Communication Development Division
William Sharp
Chief of Communication Development
Joseph Reid

The Communication Development Division was established in September 1969. To affect institutional policy realignment in the communications industry to be more responsive to the poor and other minorities.

NAME: U.S. Department of Justice
Community Relation Services

LOCATION: 500-11th Street, N.W., Room 513
Washington, D.C. 20530

TELEPHONE: (202) 739-4002

STAFF: Communication Chief, Willis A. Selden
Program Officer, Larry Still
Communication Specialist, Arthur A. Peltz
Program Assistant, Jane Redmon

Community Relation Services provides technical assistance to minorities in the fields of broadcasting, ownership, management, training, employment and programming in the communication industry.

NAME: Department of Health, Education, and Welfare
Office of Telecommunication Policy

LOCATION: 330 Independence Avenue, S.W., Room 5400
Washington, D.C. 20207

TELEPHONE: (202) 963-5941

STAFF: Director of Telecommunication Policy
Dr. Albert Horley
Assistant to the Director
Ann Erdman

The Office of Telecommunication Policy was established in July 1970 to plan policy between agencies and other departments, develop support documents for legislation, and provide background information for the Secretary of Health, Education, and Welfare in the area of telecommunication policy.

NAME: U.S. Department of Commerce
Office of Telecommunications

LOCATION: 1325 G Street, N.W.
Washington, D. C. 20230

TELEPHONE: (202) 967-4481

STAFF: Robert S. Powers
Special Assistant for Urban Telecommunications

The Office of Telecommunications was established in July 1971. The office is comprised of four divisions. In addition to assignment of radio frequencies to Federal users, long term technical and economic research, and the conduct of scientific and engineering studies, the office provides direct staff support to the Office of Telecommunications Policy. OTP is the principal advisor to the President on CATV matters.

NAME: U. S. Copyright Office

LOCATION: c/o The Library of Congress
Washington, D.C. 20540

TELEPHONE: (202) 655-4000

Register of Copyright: Abraham L. Kaminstein
Deputy Register of Copyright: George D. Cary
General Counsel: Abe A. Goldman
Chief, Reference Division: Waldo H. Moore

congressional committees

NAME: Senate Commerce Committee
Communication Subcommittee

LOCATION: c/o United States Senate
Washington, D.C. 20510

TELEPHONE: (202) 225-5115

Chairman: Warren G. Magnuson (D-Wash.)

Members:

Democrats:
Howard W. Cannon, Nevada*
Philip A. Hart, Michigan*
Vance Hartke, Indiana*
Ernest F. Hollings, South Carolina
Daniel K. Inouye, Hawaii
Russell B. Long, Louisiana*
Frank E. Moss, Utah*
John A. Pastore, Rhode Island**
William B. Spong, Jr., Virginia

Republicans:
Howard H. Baker, Jr., Tennessee*
Marlow W. Cook, Kentucky
Norris Cotton, New Hampshire
Charles E. Goodell, New York*
Robert P. Griffin, Michigan*
James B. Pearson, Kansas
Hugh Scott, Pennsylvania*

* Indicates members of Communication Subcommittee
** Indicates Chairman, Communication Subcommittee

NAME: House Commerce Committee
Communication Subcommittee

LOCATION: c/o United States House of Representatives
Washington, D.C. 20515

TELEPHONE: (202) 225-2927

Chairman: Harley O. Staggers (D-W. Va.)

Members:

Democrats:
Brock Adams, Washington
Ray Blanton, Tennessee
Bob Eckhardt, Texas
John D. Dingell, Michigan
Samuel N. Friedel, Maryland
John Jarman, Oklahoma
Peter N. Kyros, Maine
Torbert Macdonald, Massachusetts**
John E. Moss, California
John M. Murphy, New York
J. J. (Jake) Pickle, Texas
Richardson Preyer, North Carolina
Paul G. Rogers, Florida
Daniel J. Ronan, Illinois
Fred B. Rooney, Pennsylvania*
David E. Satterfield, III, Virginia
W. S. Stuckey, Georgia
Robert O. Tiernan, Rhode Island*
Lionel VanDeerlin, California*

Republicans:

Donald Brotzman, Colorado*
Clarence J. Brown, Jr., Ohio*
James T. Broyhill, North Carolina
Tim Lee Carter, Kentucky
Glenn Cunningham, Nebraska
Samuel L. Devine, Ohio
James F. Hastings, New York
James Harvey, Michigan*
Hastings Keith, Massachusetts
Dan Kuykendall, Tennessee
Ancher Nelsen, Minnesota
Joe Skubitz, Kansas
William L. Springer, Illinois
Fletcher Thompson, Georgia
G. Robert Watkins, Pennsylvania
Albert W. Watson, South Carolina

* Indicates members of Communication Subcommittee
** Indicates Chairman, Communication Subcommittee

NAME: Senate Judiciary Committee
Copyright Subcommittee

LOCATION: c/o United States Senate
Washington, D.C. 20510

Chairman: James O. Eastland (D-Miss.)

Members:

Democrats:

Birch Bayh, Indiana
Robert C. Byrd, West Virginia
Quentin N. Burdick, North Dakota*
Thomas J. Dodd, Connecticut
Sam J. Ervin, Jr., North Carolina
Philip A. Hart, Michigan*
Edward M. Kennedy, Massachusetts
John L. McClellan, Arkansas**

Republicans:

Marlow W. Cook, Kentucky
Hiram L. Fong, Hawaii
Robert P. Griffin, Michigan
Roman L. Hruska, Nebraska
Charles M. Mathias, Maryland
Hugh Scott, Pennsylvania*
Strom Thurmond, South Carolina

*Indicates members of Copyright Subcommittee
**Indicates Chairman, Copyright Subcommittee

NAME: House Judiciary Committee
Copyright Subcommittee

LOCATION: c/o United States House of Representatives
Washington, D.C. 20515

TELEPHONE: (202) 225-3927

Chairman: Emanuel Celler (D-N.Y.)

Members:

Democrats:

Jack Brooks, Texas
John Conyers, Jr., Michigan*
Harold D. Donohue, Massachusetts
John Dowdy, Texas
Joshua Eilberg, Pennsylvania
Don Edwards, California
Edwin W. Edwards, Louisiana
Michael A. Feighan, Ohio
Walter Flowers, Alabama
William L. Hungate, Missouri

* Indicates members of Copyright Subcommittee
** Indicates Chairman, Copyright Subcommittee

Andrew Jacobs, Jr., Indiana
Robert W. Kastenmeier, Wisconsin**
James R. Mann, South Carolina
Abner J. Mikva, Illinois*
Byron G. Rogers, Colorado
Peter W. Rodino, Jr., New Jersey
William F. Ryan, New York*
William L. St. Onge, Connecticut*
Jerome R. Waldie, California

Republicans:

Edward B. Biester, Jr., Pennsylvania*
R. Lawrence Coughlin, Pennsylvania
David W. Dennis, Indiana
Hamilton Fish, Jr., New York
Edward Hutchinson, Michigan*
Clark MacGregor, Minnesota*
Robert McClory, Illinois
William M. McCulloch, Ohio
Wiley Mayne, Iowa
Thomas Meskill, Connecticut
Richard H. Poff, Virginia*
Thomas F. Rallsbeck, Illinois
Charles W. Sandman, New Jersey
Henry P. Smith, New York
Charles E. Wiggins, California

* Indicates members of Copyright Subcommittee
** Indicates Chairman, Copyright Subcommittee

---

private organizations and associations

NAME: Communication Satellite Corporation (COMSAT)

LOCATION: 950 L'Enfant Plaza South, S.W.
Washington, D.C. 20024

TELEPHONE: (202) 544-6000

INFORMATION: COMSAT is chartered by an act of Congress, the Communications Satellite Act of 1962. It is a private company regulated like all other communication carrier companies. COMSAT receives revenues from leasing satellite circuits to customers. Example: Apollo 11 moon-landing was seen on TV via satellite by some 500 million people in 40 different countries.

CATV: COMSAT is competing with other major firms for the contract to build and launch a domestic satellite system. This system could transmit programs directly into CATV systems and create a fourth television network.

PRESIDENT: Joseph V. Charyk

BOARD OF DIRECTORS:

Joseph H. McConnell
Chairman of the COMSAT Board of Directors, and President, Reynolds Metals Company, Richmond, Va.

George S. Beinetti
President, Rochester Telephone Company, Rochester, N.Y.

Philip W. Buchen
Partner, Law, Buchen, Weathers, Richardson & Dutcher (Attorneys), Grand Rapids, Mich.

Joseph V. Charyk
President, Communications Satellite Corporation, Washington, D.C.

James E. Dingman
Business Consultant and former Vice Chairman of the Board, American Telephone and Telegraph Company, New York, N.Y.

Frederic G. Donner
Director and former Chairman of the Board, General Motors Corporation, and Chairman, Alfred P. Sloan Foundation, New York, N.Y. (Presidential Appointee)

William W. Hagerty, President, Drexel University, Philadelphia, Pennsylvania

Richard R. Hough, President, Long Lines Department, American Telephone and Telegraph Company, New York, N.Y.

George L. Killon, Chairman of the Board of Directors, Metro-Goldwyn-Mayer, Inc., New York, N.Y.

James McCormack, Chairman of the Board of Directors, Aerospace Corporation, and former Chairman and Chief Executive Officer, Communications Satellite Corporation, Washington, D.C.

George Meany, President, AFL-CIO, Washington, D.C. (Presidential Appointee)

Horace P. Moulton, Vice President and General Counsel, American Telephone and Telegraph Company, New York, N.Y.

Rudolph A. Peterson, Chairman of the Executive Committee and former President, Bank of America, San Francisco, Calif. (Presidential Appointee)

Bruce G. Sundlun, Partner, Amram, Hahn, Sundlun and Sandground (Attorneys), Providence, R.I., and Washington, D.C., and President, Executive Jet Aviation, Inc.

Leo D. Welch, former Chairman and Chief Executive Officer, Communications Satellite Corporation

NAME: Corporation for Public Broadcasting

LOCATION: 888-16th Street, N.W.
Washington, D.C. 20006

TELEPHONE: (202) 293-6160

INFORMATION: The Corporation for Public Broadcasting was established by Congress in 1967 as a private nonprofit corporation to promote and help finance development of noncommercial radio and television. This includes providing a national interconnection of stations and generating quality programming from many sources. CPB does not produce programs but provides grants to production centers. Its funds come from the Federal Government and private sources. Its board is appointed by the President with advice and consent of the Senate.

CATV: CPB is interested in cable television because they hope public TV will become the producers and operators for noncommercial channels on CATV systems. Additionally, CATV systems connected with satellites will provide cheaper and more efficient communication services to communities delivering educational and public service material.

BOARD OF DIRECTORS:

Chairman: Frank Pace, Jr.
President, International
Executive Service Corps
New York, New York

Vice Chairman: Dr. James R. Killian, Jr., Chairman of the Corporation, Massachusetts Institute of Technology
Cambridge, Massachusetts

Joseph A. Beirne, President
Communications Workers of America, Washington, D.C.

Robert S. Benjamin, Chairman of the Board, United Artists Corporation, New York, New York

Albert L. Cole, Vice President
Director, Reader's Digest Association
Pleasantville, New York

Michael A. Gammino, Jr., President
Columbus National Bank of Rhode Island, Providence, Rhode Island

Saul Haas, Chairman of the Board
KIRO, Incorporated, Seattle, Washington

John W. Macy, Jr., President
Chief Executive Officer

Mrs. Oveta Culp Hobby, Editor
Chairman of the Board, The Houston Post, Houston, Texas

Jack Valenti, President
Motion Picture Association of America, Incorporated
New York, New York

Joseph D. Hughes, Vice President
T. Mellon and Sons, Pittsburgh, Pennsylvania

NAME: National Association of Broadcasters

LOCATION: 1770 N Street, N.W.
Washington, D.C. 20036

TELEPHONE: (202) 293-3500

INFORMATION: NAB was established in 1922 as a nonprofit organization to promote visual broadcasting, protect its members, and encourage practices which will strengthen and maintain the broadcast industry and best serve the public.

PRESIDENT: Vincent Wasilewski

PUBLICATIONS: "Hot line" free weekly newsletter covering the broadcast business for members only.

BOARD OF DIRECTORS:

Richard L. Beesemyer, ABC-TV
New York, New York

Peter B. Kenney, NBC-TV
Washington, D.C.

William B. Lodge, CBS-TV
New York, New York

EXECUTIVE COMMITTEE:

Willard E. Walbridge, Chairman
Capitol Cities Broadcasting Corporation,
Houston, Texas

James M. Caldwell, WAVE
Louisville, Kentucky

Richard W. Chapin, Stuart Enterprises,
Lincoln, Nebraska

Grover C. Cobb, Gannett Company, Inc.,
Rochester, New York

Harold Essex, WSJS Stations, Winston-Salem, North Carolina

Hamilton Shea, Gilmore Broadcasting
Group, Harrisonburg, Virginia

NAME: National Association of Educational Broadcasters

LOCATION: 1343 Connecticut Avenue, N.W., 11th Floor
Washington, D.C. 20036

TELEPHONE: (202) 667-6000

INFORMATION: NAEB is an association established in 1934 to serve professional needs of educational radio and television. Its membership includes more than 400 noncommercial radio and television stations.

CATV: NAEB is particularly interested in cable television because of (1) possible ownership an (2) as an outlet for programming they have already developed.

PRESIDENT: William G. Harley

PUBLICATIONS: Educational Broadcasting Review. Nonmembers –$10.00; members–free; Bi-monthly newsletter–$10.00 for nonmembers and free for members.

BOARD OF DIRECTORS:

Warren A. Kraetzer, WHYY-WUHY
Philadelphia, Chairman

Karl Schmidt, University of Wisconsin,
Madison, Vice Chairman

Edmund F. Ball, Ball Corporation, Muncie, Indiana

Warren Cannon, McKinsey & Company

Floyd T. Christian, State Department of Education, Tallahassee, Florida

Myron Curry, University of North Dakota

Dr. Lark Daniel, S.E.C.A., Columbia, South Carolina

Richard Estell, Michigan State University, East Lansing, Michigan

Erling S. Jorgensen, Michigan State University, East Lansing, Michigan

Hartford N. Gunn, Jr., WGBH-TV-FM, Boston, Massachusetts

William G. Harley, NAEB, Washington, D.C.

E. William Henry, Management TV Systems, New York, New York

Harold E. Hill, University of Colorado, Boulder, Colorado

Howard D. Holst, Memphis State University, Memphis, Tennessee

Will I. Lewis, WBUR, Boston, Massachusetts

Leonard Marks, Cohn & Marks, Washington, D.C.

Wanda Mitchell, Evanston Twp. Schools, Evanston, Illinois

Dr. Carroll W. Newsom, New Fairfield, Connecticut

Layhmond Robinson, National Urban League, New York, New York

Elvis Stahr, National Audubon Society, New York, New York

Loren B. Stone, University of Washington, Seattle, Washington

---

NAME: National Cable Television Association, Inc.

LOCATION: 918-16th Street, N.W., Suite 800
Washington, D.C. 20006

TELEPHONE: (202) 466-8111

INFORMATION: NCTA was established in 1952 as a national trade organization representing the CATV industry before the Federal Communications Commission and state regulatory bodies. Included in its active membership are 1150 operating CATV systems and 210 associate members. Associate members are manufacturers and suppliers of CATV components, CATV brokerage and consulting firms, financial and other organizations having an interest in the CATV industry.

STAFF:

(vacant), President

Larry D. Bowin, Assistant to the President

Wally Briscoe, Managing Director

Gary L. Christensen, General Counsel

Charles S. Walsh, Assistant General Counsel

Herbert A. Jolovitz, Director of Government Relations

G. Norman Penwell, Director of Engineering

John K. Lady, Director of Research

Don Witheridge, Director, Public Relations Department

Donald "Gene" Burton, Director of Membership Services

Dave Roudybush, Controller

Rochelle Nezin, Membership Secretary

(Officers)

John Gwin, National Chairman
Cox Cablevision, Inc.
Robinson, Illinois

William Bresnan, Vice-Chairman
Teleprompter Corporation
San Francisco, California

F. Gordon Fuqua, Secretary
Electra Communications
Charlotte, North Carolina

Glen Scallorn, Treasurer
Communication Properties
Del Rio, Texas

Ralph Demgen, Immediate Past National Chairman
Willmar, Minnesota

PUBLICATIONS:

NCTA Films and Printed Material: (Note: Unless otherwise specified, these are available to members only.)

"CATV: A Response to Public Demand," 20-min., 16mm sound color film about the CATV industry. Purchase price, $250; week's rental, $15; 2-week's $25.

"Tune in on the Wonderful, Colorful, Colorful World of Cable TV." 12-pg. 4-color booklet about the growth and development of CATV. Price, 8 cents each to 999; 7.5 cents each 1,000-4,999;7 cents each 5,000-9,999; 6.5 cents each 10,000 or more. Also available to nonmembers in quantities up to 50 copies.

---

political action committee of cable television

Cable television operators have been making campaign contributions through the Political Action Committee of Cable Television. These contributions have been given to members of the House, Senate, Judiciary and Commerce Committees, which handle most legislation concerning the CATV industry.

PACCT was formed in June 1969 by Martin F. Malarkey, co-founder of the NCTA in 1951 and president of Malarkey, Taylor & Associates of Washington, D.C., the nation's largest CATV consulting and engineering firm.

The Committee's contribution through 1970 totaled $21,250. All recipients are listed below with the amount of contribution each received.

All members listed were reelected. Three contributions were made in 1969; all others were made in 1970, the election year.

Senators

Bayh, D-Ind.	$ 500
Burdick, N.-N.D.	500
Byrd, D-W. Va.	500
Cannon, D-Nev.	1,500
Fong, R-Hawaii	1,300
Hart, D-Mich.	1,000
Hartke, D-Ind.	1,500
Jackson, D-Wash.	500
Moss, D-Utah	1,500
Prouty, R-Vt.	500
Scott, R-Pa.	1,700

Representatives

Brotzman, R-Colo.	$ 500
Brown, R-Ohio	250
Broyhill, R-N.C.	700
Celler, D-N.Y.	500
Clark, D-Mich.	100
Conyers, D-Mich.	200
Donohue, D-Mass.	200
Edwards, D-Calif.	200
Harvey, R-Mich.	200
Hastings, R-N.Y.	150
Jarman, D-Oka.	250
Keith, R-Mass.	100
Kyros, D-Maine	400
Long, D-Md.	100
Macdonald, D-Mass.	1,500
McCulloch, R-Ohio	$ 500
Mikva, D-Ill.	150
Murphy, D-N.Y.	250
Preyer, D-N.C.	400
Quillen, R-Tenn.	100
Rodino, D-N.J.	250
Rooney, D-Pa.	1,000
Stratton, D-N.Y.	350
Tiernan, D-R.I.	500
Van Deerlin, D-Calif.	1,000
Yatron, D-Pa.	100

Nonincumbents

Chavez, D-N.M.	$ 200
Tunney, D-Calif.	100

SOURCE: Political Action Committee of Cable Television c/o Malarky, Taylor & Associates Washington, D.C.

research and demonstration projects

Studies and demonstration projects underway or planned should be carefully monitored by community organizations to assure that their interests and needs are properly included and considered. Minority professionals in law, engineering, education and media productions in local communities should be contacted to assist in the monitoring. Community development groups should seriously consider establishing their own study and research projects in the areas of system design, costs, programming, financing, and marketing. Community Action Agencies, Model Cities Programs, Urban Renewal Agencies, and Community Development Corporations are possible sponsors for these projects.

The majority of the research and demonstration activities initiated over the past two years are listed here. The research findings in each case could be very useful. Project directors or the funding agency should be contacted for copies of the funding proposal and the research findings if the project has been completed.

NAME: Office of Telecommunication Policy

PROJECT DIRECTOR: Norman Pinwell

LOCATION: Malarkey, Taylor and Associates
1225 Connecticut Avenue, N.W.
Washington, D.C. 20036

TELEPHONE: (202) 395-3782

PROJECT DESCRIPTION: A study by Malarkey and Taylor CATV consultants, to identify new broadband services which might be offered on a pilot basis to test their social benefits, public acceptance and economic viability of such services. An example might take the form of connecting two diverse communities with two-way cable capabilities to serve as a test bed for controlled evaluation of alternative services in education, recreation, health, banking, public administration and public safety. Completion date for the study is late November 1971.

NAME: Institute of Film and Television. New York University School of the Arts

DIRECTOR: George Stoney

LOCATION: South Building
51 West 4th Street
New York, New York

TELEPHONE: (212) 598-3703

PROJECT DESCRIPTIONS: *NYU Media Club*, a community film project, producing video tapes covering all subjects and shown daily. Funded by the New York State Council of the Arts.
Videotex, a community video access center. A model for the development of community video centers. Over 100 people were trained in four months. Funded by New York University ($10,000).
Alternate Media Center, established as a prototype for the development of local origination programs for CATV. (212) 598-3339. Funded by the Markle Foundation ($250,000).

NAME: RAND Corporation

LOCATION: 1700 Main Street
Santa Monica, California 90406

TELEPHONE: (213) 393-0411

FUNDS: Funded by the John and Mary Markle Foundation—$500,000 grant for three years.

PROJECT DESCRIPTION: RAND Corporation was funded to examine CATV local origination, signal importation, over-the-air broadcasting impact, regulatory functions at all government levels and over all regulatory policies for CATV.

PUBLICATIONS FROM ABOVE GRANT:

Cable Television: The Problem of Local Monopoly. The RAND Corporation. May 1970.

Potential Impact of Cable Growth in Television Broadcasting. The RAND Corporation. October 1970.

Cable Television and the Question of Protecting Local Broadcasting. The RAND Corporation. October 1970.

Cable Television: Opportunities and Problems in Local Program Origination. The RAND Corporation. September 1970.

Telecommunications in Urban Development. The RAND Corporation. July 1969.

Communications Goals for Los Angeles—A Working Paper for the Los Angeles Goals Program. The RAND Corporation. June 1968.

NAME: National Academy of Engineering

DIRECTOR: Peter C. Goldmark, Sr.

LOCATION: 2101 Constitution Avenue, N.W.
Washington, D.C.

TELEPHONE: (202) 393-8100

FUNDS: Funded by the Departments of Housing and Urban Development, Commerce and Justice.

PROJECT DESCRIPTION: The Telecommunications Committee of NAE was funded to study better application of telecommunications technology to improve living conditions and stimulate regional development. Four pilot projects were suggested in the Academy's report, *Communications Technology for Urban Improvement*, June 1971:

> Design and demonstrate model Community Information Center, serviced with modern video, facsimile, and telephone systems, to provide improved city services to the citizens.
>
> Explore the effectiveness of various forms of two-way instructional television in improving the quality and distribution of educational services among the urban population.
>
> Demonstrate the provision of community information retrieval services to the school and home over cable television with limited subscriber response capacity and interactive terminals.
>
> . . .explore the potential for distribution of CAI (computer-assisted instruction) services over cable television with limited subscriber feedback capacity.

NAME: Foundation 70

DIRECTORS: Frank White, Jeffrey Stamps, Jessica Lipnack

LOCATION: 55 Chapel Street
Newton, Massachusetts 02160

TELEPHONE: (617) 237-4977

FUNDS: Grant received from the John and Mary Markle Foundation

PROJECT DESCRIPTION: Foundation 70 was funded to develop and disseminate information on CATV to the public. It conducts cable communications research for community groups and public institutions in Boston and the State of Massachusetts.

NAME: Mitre Corporation/Washington Operations

DIRECTOR: William F. Mason

LOCATION: Westgate Research Park
McLean, Virginia 22101

TELEPHONE: (202) 983-3500

FUNDS: John and Mary Markle Foundation ($250,000).

PROJECT DESCRIPTION: Mitre Corporation was funded to study the application of CATV to an urban area, specifically for the District of Columbia. Objectives include a system design for the entire city, plan of installation, projection for future growth, and development of innovative community uses for cable communication. System design is comprised of a cablecasting service for the public and a point-to-point system for possible federal and local government use. Completion date: November 1971.

NAME: RAND Study—Dayton

DIRECTOR: Dr. Leland L. Johnson

LOCATION: RAND
1700 Main Street
Santa Monica, California

Bonnie Macaulay
Council of Governments
Cable Subcommittee
Dayton, Ohio

Stuart Sucherman
Ford Foundation
320 East 43rd Street
New York, New York

TELEPHONE: (213) 393-0411

FUNDS: Funded jointly by the Kettering and Ford Foundations ($80,000).

PROJECT DESCRIPTION: The objectives of the Dayton Study conducted by the RAND Corporation include establishing guidelines for selecting franchise applicants and determining cost and benefits related to various ownership and design configurations, including a regional system. The study will also explore ownership structure for a system in Dayton; determining the impact of a cable system on local television; kinds of services which could be provided; and the types of experimental projects to test new programming and technology. Completion date: December 1971.

NAME:	Bedford-Stuyvesant Restoration Corporation
PROJECT DIRECTOR:	Barry Lemieux
LOCATION:	1368 Fulton Street Brooklyn, New York 11216
TELEPHONE:	(212) 636-1100
FUNDS:	Grants received from the Sloan Foundation and individuals.
PROJECT DESCRIPTION:	A feasibility study by Malarkey and Taylor, CATV consultants/engineers, to determine the optimal CATV system, cost and methods of financing for the Bedford-Stuyvesant community in Brooklyn, New York—an area involving 500,000 people and 136,000 households. Basic data expected to be completed by fall 1971.

NAME:	Columbia University
DIRECTOR:	Dr. Amitai Etzioni
LOCATION:	Bureau of Applied Social Research 423 West 18th Street New York, New York
TELEPHONE:	(212) 870-2011/288-3693
FUNDS:	Funded by the National Science Foundation ($124,000).
PROJECT DESCRIPTION:	A one-year grant to the Center for Policy Research to study the uses of CATV as a means of community communication and decision-making. CATV and Network Television will also be compared to see how they duplicate or complement each other. Completion date is May 31, 1972.

NAME:	Sloan Commission on Cable Communications
DIRECTOR:	Paul L. Laskin
LOCATION:	105 Madison Avenue New York, New York 10016
TELEPHONE:	(212) 684-1800
FUNDS:	Funded by the Alfred P. Sloan Foundation ($500,000).
PROJECT DESCRIPTION:	The Commission was established to determine whether the social and economic needs of urban areas can be met by cable television, to make a technological assessment of cable, and to provide information to city governments about franchising. Publication of findings and report: December 1971.

information and technical assistance

In addition to hardware and equipment manufacturers and consulting firms there are numerous sources of information and technical assistance for community groups interested in cable communications. The organizations included in the following represent a broad spectrum of knowledge and experience in broadcasting and related media fields. Community groups are advised to use these lists selectively to develop useful contacts, and seek information and technical assistance. Local media specialists, including those on the staffs of college and university-based video and radio centers should not be overlooked.

NAME:	Black Efforts for Soul in Television (BEST)
LOCATION:	1014 North Carolina Avenue, S.E. Washington, D.C. 20003
PHONE:	(202) 547-1286
NATIONAL COORDINATOR:	Bill Wright
PUBLICATIONS:	"Guide to Community Demands in License Challenges" "Law About Television" (all free)
INFORMATION:	BEST was established in September 1969 to advise individuals and community groups of their rights in the broadcast media. BEST also provides technical assistance in license challenges, programming complaints or employment discrimination.

NAME:	Citizens Communication Center
LOCATION:	1812 N Street, N.W. Washington, D.C. 20036
PHONE:	(202) 296-4238
DIRECTOR:	Albert H. Kramer
PUBLICATIONS:	"Primer on Citizen Access to the Federal Communications Commission" (free)
INFORMATION:	The Citizen Communication Center was established in August 1969 to provide the public with legal advice and representation before the FCC and Federal Courts on communication issues of social importance.

NAME:	National Mexican-American Anti-Defamation Committee (NMAADC)
LOCATION:	1346 Connecticut Avenue, N.W. Washington, D.C. 20036
PHONE:	(202) 833-2667
DIRECTOR:	Domingo Nick Reyes
INFORMATION:	NMAAC was established to fight the media's frequently false sterotype of Spanish-speaking Americans. The organization's focus is on citizen action in broadcasting, increasing training and employment opportunities, and ascertaining community needs.

NAME:	Urban Law Institute
LOCATION:	1145-19th Street, N.W. Washington, D.C. 20036
PHONE:	(202) 833-1700
DIRECTOR:	Jean Camper Cahn
INFORMATION:	The Urban Law Institute aids communities in their efforts to develop more responsive service from the broadcast industry. Legal advice is given on license challenges, programming complaints, and ordinance requirements for cities to insure greater minority access into programming content and employment.

NAME:	NAACP Legal Defense and Educational Fund (LDF)
LOCATION:	10 Columbus Circle New York, New York 10019
PHONE:	(212) 586-8397
INFORMATION:	The Division of Legal Information and Community Service of LDF assists in organizing community groups, conducting workshops on analyzing renewal applications, monitoring media services, and assisting in the preparation of formal and informal complaints.

NAME:	PUBLICABLE
LOCATION:	c/o Dr. Harold Wigren National Education Association 1201-16th Street, N.W. Washington, D.C. 20036
PHONE:	(202) 833-4120
DIRECTOR:	Acting Director, Dr. Harold E. Wigren
INFORMATION:	PUBLICABLE was established in July 1971 in an attempt to stimulate public interest in cable television. The organization seeks to identify issues in cable communication and provide information and assistance to varied public interest groups. PUBLICABLE is represented by individuals from many national educational, public service and community organizations.

NAME:	Office of Communication, United Church of Christ
LOCATION:	289 Park Avenue South New York, New York 10010
PHONE:	(212) 475-2121, ext. 266
DIRECTOR:	Dr. Everett C. Parker
PUBLICATIONS:	Lawyer's Sourcebook for Citizen Action in Radio and TV (free) (Available Summer 1972) *A Guide to Understanding Broadcast License Application* by Dr. Ralph M. Jennings ($1.00 per copy) In Defense of Fairness (free) Racial Justice in Broadcasting (free) *Guide to Citizen Action in Radio and TV* by Marsha O'Bannon Prowitt (free)

Cable Television A manual by John Wicklein and Monroe Price, to be published Summer 1972, 200 pp., $2.95. Community groups may obtain a single copy free on request by writing to the preceding address.

INFORMATION: The Office of Communication was established to assist minorities in programming and employment in the broadcast field. It offers community groups advice by mail or telephone, provides staff assistance in the field and legal assistance in preparing petitions to deny license renewals.

NAME: Stern Community Law Firm

LOCATION: 2005 L Street, N.W.
Washington, D.C. 20036

PHONE: (202) 659-8132

DIRECTOR: Tracy Weston

INFORMATION: Stern Community Law Firm offers legal advice in broadcasting and CATV matters. Emphasis is on fairness doctrine, right of access, and other First Amendment issues.

NAME: Urban Communications Group

LOCATION: 1730 M Street, N.W.
Washington, D.C. 20036

PHONE: (202) 223-4916

DIRECTOR: Ted Ledbetter

INFORMATION: Urban Communications Group (UCG) is the sole black consulting firm with technical expertise in cable television. UCG also provides consultant services to federal, state and local agencies, community organizations, business, foundations and others who want to increase public, diversified ownership, and greater responsiveness to minority concerns in the broadcast media community.

NAME: Chamba Productions, Inc.

LOCATION: 630 Ninth Avenue, Room 900
New York, New York 10036

PHONE: (212) 489-0790

PRESIDENT: St. Clair Bourne

INFORMATION: Chamba Productions is composed of four experienced film producers who make movies of all types. It is currently doing a movie for the Brooklyn Museum in New York City. No training services are available now. Hope to have some available in the future.

NAME: Bill Fields Media, Ltd.

LOCATION: 2633 Franklin Avenue
St. Louis, Missouri 63106

PHONE: (314) 533-2576

PRESIDENT: Bill Fields

INFORMATION: Bill Fields Media, established in 1967, produces documentary films with social implications, also educational films and advertisements. "The Corner" (on TV) 1965 Emmy Nominee; "What's a Man Worth?" (1968 Emmy).

NAME: Jemmin Productions, Inc.

LOCATION: 1800 North Highland Avenue
Hollywood, California 90028

PHONE: (213) 461-4087

PRODUCER: Del Shields/Bill Cosby

INFORMATION: Jemmin Productions is owned by Bill Cosby and produces his TV series, and his latest movie "Man and Boy."

NAME: Jymie Productions, Inc.

LOCATION: 866 Sterling Place
Brooklyn, New York 11216

PHONE: (212) 467-6416

PRESIDENT: James M. Mannas, Jr.

INFORMATION: Jymie productions was established in 1965. Expertise covers film production and distribution, still photography, graphics, documentary and animated films. Training given by professionals in motion picture techniques.

NAME: Ken Snyder Enterprises

LOCATION: 6335 Homewood Avenue
Hollywood, California 90028

PHONE: (213) 462-6758

PRESIDENT: Kenneth Snyder

INFORMATION: Ken Snyder Enterprises' expertise covers commercial, industrial, educational films. Customers: U.S. Job Corps; Institute of Mental Health on Dope Addiction.

NAME: Lencia Production Associates, Inc.

LOCATION: 305 West 19th Street
New York, New York 10011

PHONE: (212) 924-9517

PRESIDENT: Lenny Lencia

INFORMATION: Lencia Productions was established in January 1971. Expertise covers writing, developing and producing motion pictures, single or double systems, still photography or animations.

NAME: Oscar Production, Inc.

LOCATION: 727 23rd Avenue, South Dearborn
Seattle, Washington 98144

PHONE: (206) EA. 4-9440

PRESIDENT: Nat Long

INFORMATION: Oscar Production was established in 1969. Expertise covers motion picture production, still photography. Training is offered using 16 mm cameras, video tape and porta-pack equipment. Currently producing a public service TV show titled "Comma."

NAME: Reels and Reality Film Company

LOCATION: 9340 South Lafayette Avenue
Chicago, Illinois 60620

PHONE: Mr. Grant

INFORMATION: Reels and Reality was established in 1968. They produce documentary, training, industrial and TV commercials using Aries and Eclair cameras with Nagra tape recorders. Customers: HEW, HUD, SBA, Operation Breadbasket.

NAME: Silhouetts in Courage, Inc.

LOCATION: 22 East 40th Street
New York, New York 10016

PHONE: (212) 265-1240

PRESIDENT: Charles Jones

INFORMATION: Silhouetts in Courage, established in 1968, produces educational and commercial TV films: Doo Dat Productions, also music publishing: "Stomp-Town" label. Customer: Scanlan Press (NYC), Mastertone studios, (NYC), Audio Matrix (NYC).

NAME: Summit Productions

LOCATION: 538 East Alameda Avenue
Denver, Colorado 80209

PHONE: (303) 744-1319

PRESIDENT: Roger Brown

INFORMATION: Summit Productions was established in 1961. Expertise covers motion pictures, still photography, commercials, and animations. No training programs are offered. Summit Production, an award-winning film company, is currently doing films for Head Start and Manpower programs.

NAME: Tee Collins, Inc.

LOCATION: 2 West 45th Street, Suite 800
New York, New York 10036

PHONE: (212) 972-1820

PRESIDENT: Tee Collins

INFORMATION: Tee Collins, Inc. produces animated slides, film strips and educational films. Originator of Sesame Street's "Wanda the Witch" and "The Seal." Customer: Doubleday Multimedia and Children's TV Workshop.

NAME: Third World Productions, Inc.

LOCATION: 62 West 45th Street
New York, New York

PHONE: (212) 972-9300

PRESIDENT: Ossie Davis

INFORMATION: Third World Cinema is a feature film motion picture company. Currently shooting the Billie Holiday Story. On the job training is available covering all phases of the industry. Plans are underway to establish the New Cinema Institute to train blacks in film making.

NAME:	Trans Oceanic Production, Inc.
LOCATION:	5352 West Pico Boulevard Los Angeles, California 90019
PHONE:	(213) 296-3770
PRESIDENT:	Charles Greene
INFORMATION:	Trans Oceanic Production creates and produces films from script writing to finished print. They do documentary, story, training, industrial or educational films, cartoon animation, commercials and still photography.

NAME:	William Greaves Production, Inc.
LOCATION:	254 West 54th Street New York, New York 10019
PHONE:	(212) 586-7710
PRESIDENT:	William Greaves
INFORMATION:	William Greaves Productions has done a number of documentaries for government agencies. Expertise also covers script writing, industrial or educational films, and commercials. No training program has been developed.

NAME:	Zebra Associates, Inc.
LOCATION:	1180 Avenue of the Americas Suite 1703 New York, New York 10036
PHONE:	(212) 586-2120
PRESIDENT:	Raymond A. League
INFORMATION:	Zebra Associates was established in May 1969. Expertise covers television and film production as well as a full range of advertising services.

NAME:	Black Academy of Arts and Letters
LOCATION:	475 Riverside Drive New York, New York 10027
PHONE:	(212) 663-4740
PRESIDENT:	Dr. C. Eric Lincoln
INFORMATION:	The year old Academy of Arts and Letters is composed of members from the arts and sciences, dance, drama, poetry, social sciences. Operating expenses are covered by a three-year grant from the Twentieth Century Fund. The purpose of the organization is to promote more activity in black arts, through awards, fellowships, and cash prizes. The awards cover authors, historians, persons in drama, cinema, poetry and other related fields.

NAME:	Black Journal
LOCATION:	10 Columbus Circle New York, New York 10019
PHONE:	(212) 262-5565
PRESIDENT:	Tony Brown
INFORMATION:	Funded by Ford Foundation and public funds provided by National Educational Television. Black Journal is a black TV news journal produced for and about black people.

NAME:	Community Film Workshop Council
LOCATION:	112 West 31st Street New York, New York 10001
PHONE:	(212) 239-4420
DIRECTOR:	Cliff Frazier
INFORMATION:	CFW was formed in 1968 by the American Film Institute with a seed grant of $50,000 to help New York film organizations recruit minority-group members. OEO funding supports two training programs: one was the establishment of 10 community film workshops throughout the country; the other was the training of 30 new cameramen in two six-week (now nine weeks) cycles. Entry requirements are an annual income

less than $1,600 net, some proficiency in photography and willingness to relocate in other parts of the country. Upon completion of the nine-week course, CFW actively seeks employment for its graduates.

NAME: Mafundi Institute

LOCATION: 1827 East 103rd Street
Los Angeles, California 90002

PHONE: (213) 564-4496

PRESIDENT: Dr. J. Alfred Cannon

INFORMATION: In the summer of 1966, The Communications Foundation and the Kettering Foundation initiated a project in film-making for high school dropouts in the Watts district of Los Angeles. The results were so startling and the continuing demand for advanced training so great, that the Kettering Foundation and the Council on Church and Race granted seed money for a permanent program of artistic training under community direction and control. The Mafundi Institude (Swahili for "creative artisan") was the result and became a center for training in the performing arts.

In 1967, film-making, drama, and motion graphics programs were established. By 1968, the Institute was serving some two hundred and fifty students and conducting classes in the above plus courses in literature, dance, fine arts, music, voice, fencing and modeling. The professional instructors included Don Mitchell, Raymond Burr, Marie Bryant, Marge Champion, William Marshall, and Raymond St. Jacques. In 1969, a professional radio announcer course was added to the curriculum and radio station KPFK established its Watts Bureau at Mafundi.

NAME: Fide's House Communications Workshop

LOCATION: 1554-8th Street, N.W.
Washington, D.C. 20009

PHONE: (202) 265-2130

DIRECTOR: Vernard Gray

INFORMATION: A Community Communications Center is being established to involve the D.C. community in still photography, video taping, and mobile video, and to develop trained personnel for local video programming production.

NAME:	New Thing Art and Architecture Center, Inc.
LOCATION:	1811 Columbia Road, N.W. Washington, D.C. 20009
PHONE:	(202) 332-1811
DIRECTOR:	Topper Carew
INFORMATION:	New Thing was designed to improve blacks' self-image. Activities include a learning center, children's programs, entertainment and economic development. Training is available in motion picture production, photography, one-half inch video tape productions. A production company, New Thing Flick Company, has also been established.

black owned radio stations

ALABAMA

WEUP Radio
2609 Jordan Lane, Huntsville 35806
(205) 536-0713 Leroy Garrett, Owner
5000 watts. Greener, Haiken & Sears rating. Owned by Garrett Broadcasting, Inc. (estab '58, empl 12, sales $90,000)

GEORGIA

WRDW Radio
1480 Eisenhower Drive, Augusta
(404) 738-2513 James Brown, Owner
5000 watts. Owned by JB Broadcasting. (estab '69, empl 19)

ILLINOIS

WMPP Radio
1000 Lincoln Highway, East Chicago 60411
(312) 785-0262 Charles J. Pinckard, Owner
50-mile radius includes Gary and Chicago. (estab '62, empl 18)

INDIANA

WTLC-FM Radio
1734 Villa Avenue, Indianapolis 46203
(317) 784-4471 Frank Lloyd, President
Owned by Calojay Enterprises.

MARYLAND

WEBB Radio
Clifton & Dennison, Baltimore
(301) 947-1245 James Brown, Owner
5000 watts directional. Owned by JB Broadcasting. (estab '69, empl 18)

MICHIGAN

WCHB-AM and WCHD-FM Radio
32790 Henry Ruff Road, Inkster 48141
(313) 278-1440 Dr. Haley Bell, President
WCHB-AM transmits 1000 watts directionally in metro Detroit, Wayne County, and Toledo, reaching 5 million with soul music. WCHD-FM has 75 mile radius and jazz music. Owned by Bell Broadcasting, Inc. (estab '56, empl 40)

WGPR-FM Radio
2101 Gratiot Street, Detroit 48207
(313) 961-8833 William V. Banks, President

WWWS-FM Radio
2721 South Washington, Saginaw 48607
(517) 752-7166 Ch. Grazen, Director
400,000 people in 70-mile radius to Lansing, Flint, and Pontiac. 5000 watts effective power. Owned by Clark Broadcasting. (estb '69, empl 14)

MISSISSIPPI

WORV Radio
604 Jussia Avenue, Hattiesburg 39401
(601) 582-7013 Vernon Floyd, Owner

MISSOURI

KPRS-AM and KPRS-FM Radio
2301 Grand Avenue, Kansas City 64108
(816) 471-2100 Andrew Carter, President

KWK Radio
500 Terminal Building, St. Louis 63147
(314) 868-6440 Dr. Robert Bass, Partner
Owned by Bell Broadcasting Corporation

NORTH CAROLINA

WVOE Radio
Route 2, Box 124-A, Chadbourn 28431
(191) 654-3991 Dr. G. W. Carnes, President
Owned by WVOE Enterprises. (estab '62, empl 9)

OHIO

WABQ Radio
2644 St. Clair Avenue East, Cleveland 44114
(216) 241-7555

TENNESSEE

WJBE Radio
2108 Prosser Road, P.O. 281, Knoxville
(615) 546-2210 James Brown, Owner
1000 watts. Pulse ratings. Owned by JB Broadcasting. General Rank 3. (estb '68, empl 12)

FM stations located on black college campuses.

WCSU-FM
Central State University
Wilberforce, Ohio

WHOV-FM
Hampton Institute
Hampton, Virginia

WSHA-FM
Shaw University
Raleigh, North Carolina

minority cable companies

NAME:	District Cable Vision, Inc.
LOCATION:	1000-16th Street, N.W. Washington, D.C. 20036
PHONE:	(202) 737-5545
PRESIDENT:	E. Richard Brown
INFORMATION:	District Cable was established in 1970 and is currently applying for a District of Columbia franchise to serve one segment of the predominantly black population in the city.

NAME:	Focus Cable, Inc.
LOCATION:	2816 Haven Court Building Oakland, California
PHONE:	(415) 635-1162
PRESIDENT:	C. J. Patterson
INFORMATION:	Focus was the first predominantly black group to acquire a major CATV franchise. It is also the first joint venture between Teleprompter Corporation, the largest CATV system owner, and a black group in a large black community. The system is still in the construction stage, completion is expected in 1973. There are 151,000 potential subscribers who will receive two types of service: 1-12 channels for $1.70 per month or 38 channels for $6.15 per month.

NAME: San Diego Cable

LOCATION: 5444 Roswell Street
San Diego, California 92113

PHONE: (714) 263-4441

PRESIDENT: Chuck Johnson

INFORMATION: San Diego Cable Television, Channel 7, began cablecasting a variety of general interest and community oriented programs to 55,000 households wired to the Mission Cable system in Oct. 1970. Mission Cable is owned by the second largest multiple system owner, Cox Cable. (Further reference to SDC-TV's local origination can be found in Section I, "Local Origination" by Ted Ledbetter.)

NAME: West Essex Cable Television, Inc. (A subsidiary of Essex Cable)

LOCATION: P.O. Box 844
Newark, New Jersey 07101

PHONE: (201) 675-4312

PRESIDENT: Edward B. Lloyd

INFORMATION: West Essex Cable was established in 1967 and is currently designing a system for 4,000 households and 17,000 people in West Orange, New Jersey. Ed Lloyd, who is black and also President of Essex Cable, the parent organization, is competing for franchises in North Little Rock, Arkansas, parts of Brooklyn, New York, and Detroit, Michigan.

NAME: Watts Communications Bureau, Inc. (Subsidiary of Mafundi Institute)

LOCATION: 1827 East 103rd Street
Los Angeles, California 90002

PHONE: (213) 564-4496

ACTING DIRECTOR: Don Bushnell

INFORMATION: The Watts Communications Bureau has submitted an application to the City of Los Angeles for a franchise to install and operate a cable television system in the Watts area of Los Angeles. The Los Angeles Board of Utilities has notified the Bureau that it is a viable and responsible corporate agent for the franchise.

The Bureau is planning a bi-directional or two-way system with twenty-four forward channels and six channels for reverse direction signals.

The franchise area will comprise approximately fifty-nine square miles. The population of the area is 799,155.

publications and periodicals

One of the most effective means of finding out "what's happening" in and with cable industry is through trade journals and publications. The periodicals and publications listed are reliable and timely in their coverage of important policy actions, mergers, franchise acquisitions and changes in systems technology.

It is recommended that community groups acquire the following publications and manuals as the core of their CATV reference library.

Publication	Annual Subscription Cost
Black Communicator	$ 5.00
CATV Source Book	8.00
CATV Magazine (includes CATV Systems Directory and Map Service and CATV Directory of Equipment, Services and Manufacturers)	33.00
TV Communication Magazine	10.00
Television Factbook	35.00
CATV Station Coverage Atlas and 35-Mile Zone Maps	19.50
CATV Accounting Manual	10.00
CATV Cash Flow Projection	2.50
Total	$123.00

Accounting Manual for Cable Televison

National Cable Television Association
918-16th Street, N.W., Suite 800
Washington, D.C. 20006
(202) 466-8111

$10.00

Prepared by the National Cable Television Association, Budget and Audit Committee and Arthur Anderson & Company, offers an excellent guide to the types of accounts which may be needed in operating a system. The following is a partial listing of the types of materials covered: assets, liabilities, stockholders' equity, operating income, operating expenses, other income and other expense provision for federal and state income taxes, and extraordinary items.

Black Communicator

Urban Communications Group
1730 M Street, N.W., Suite 405
Washington, D.C. 20036

Published Monthly
$5.00 individual subscription
$15.00 organization subscription

Special focus on CATV and minority interests and issues. Also, general information on public news, in broadcasting, media monopolies, license challenges, and federal agency actions.

CATV Magazine

TV Communications
1900 WestYale
Englewood, Colorado
(303) 761-3770

Published Weekly
$33.00/year*

Weekly news magazine for system owners, management executives and all those interested in cable television. Detailed reports of news developments from the CATV Magazine Washington Bureau, franchise activity and grants, construction progress reports, financial and personality reports, new equipment and services, plus hard-hitting editorials are a part of each issue.

*Includes the following annual publications: CATV Systems Directory and Map Service
CATV Directory of Equipment Services and Manufacturers

Television Digest

2025 Eye Street, N.W.
Washington, D.C. 20006
(202) 965-1985

Published Weekly
$137.00 individual
$123.00 group rate

Weekly trade journal covering television and CATV, all agencies public and private with interest in broadcasting. Consumer electronics items are also covered, including video cassette models, rating and latest styles.

Television Factbook

Published by:
Television Digest
2025 Eye Street, N.W.
Washington, D.C. 20006
(202) 965-1985

Published Annaully
$35.00

Contains reliable reference data in the following areas:

- Abbreviations & Initials Commonly Used in TV
- Advertising Agencies
- Allocations, TV Channel
- Applications, TV
- Associations, Groups & Other Organizations
- Attorneys Directory
- CATV
  - Broadcaster Ownership
  - Canadian Systems
  - Equipment Manufacturers
  - Group Ownership
  - N.C.T.A.
  - Statistics
  - U.S. Systems
- College & Universities Offering TV-Radio Degrees
- Communications Satellite Corporation (COMSAT)
- Community Antenna TV Systems (See CATV)
- Congressional Committees
- Consulting Engineers
- CPs & Applications, TV
- Educational TV Stations
- Electronic Industries Association
- European Broadcasting Union
- FCC Directory
- Group TV Station Ownership
- Importers of TVs, Radios & Phonos
- Instructional TV
- International TV Directory
- Labor Unions & Guilds
- Management & Technical Consulting Services Manufacturers
  - CATV Equipment
  - Receivers
  - Telecasting Equipment
- Market & Audience Research Organizations
- Market Rankings, TV
- National Association of Broadcasters
- Networks, TV
- Program Sources & Services
- Publications in TV & Related Fields
- Representatives, TV Sales
- Sales & Transfers
- Statistics
- Television Stations
  - Canadian
  - Educational
  - Group Ownership
  - International
  - Studio & Mobile Equipment
  - U.S.
- U.S. Information Agency

CATV Directory of Equipment, Services and Manufacturers

Published by:
TV Communications
1900 West Yale
Englewood, Colorado 80110
(303) 761-3770

Published Annually
$6.95—available without additional charge to subscribers to CATV magazine.

Contents of this directory are compiled from material furnished by manufacturers and suppliers serving the cable television industry. The 1971 edition covers the following topics:

Antennas, Towers and Head-End Buildings
Cables, Connectors, Fittings and Accessories
CATV Professional Services
CATV Test Equipment
Construction Materials, Tools and Equipment
Distribution Equipment-Amplifiers and Passive Devices
Head-End Electronic Equipment
Local Origination Equipment and Accessories
Microwave Antennas and Equipment

CATV Station Coverage Atlas and 35 Mile Zone Maps

Published by:
Television Digest
2025 Eye Street, N.W.
Washington, D.C. 20006
(202) 965-1985

Published Annually
$19.50

Contains reliable reference data in the following areas:

ARB Market Rankings, Alphabetically
ARB Market Rankings, By Size
Amendments to FCC CATV Rules
Call Letters of U.S. Stations
CATV and Bell System Coordinators
CATV Industry Statistics
Copyright Office
Directory of Translator Stations
Directory of CATV Equipment Manufacturers
Directory of Microwave Serving CATV Systems
Directory of CATV Communities Not Shown on Maps
Directory of NCTA Regional and State Associations
FCC Cable Television Bureau
Map of Microwaves Serving CATV Systems
NEW FCC CATV Rules and Proposed Rules
Proposed FCC Rules Issued December 13, 1968
Proposed FCC Rules Issued August 6, 1971
State Maps and Indexes, Grade B Contours
State Maps, Grade A Contours and 35 Mile Zones

CATV Systems Directory and Map Service

Published by:
TV Communications
1900 West Yale
Englewood, Colorado 80110
(303) 761-3770

Published Annually
$8.95—available without additional charge to subscribers to CATV magazine.

Contents of this directory are compiled from materials furnished by Federal Agencies, cable system operators, and from research by the directory editors. Factual information is provided on the following topics:

Canadian CATV System Listings, by Community Served
CATV Map Section-System Locations by State
CATV Regulations—A Summary of the FCC's Rules
CATV Operators' Association Listings
FCC-Approved Top 100 Market List
Guide to Federal Agencies and Committees of Congress
Multiple System Operator Listings
U.S. CATV System Listings, Individually by Community Served

Broadcasting Magazine

1735 DeSales Street, N.W.
Washington, D.C. 20036
(202) 638-1022

Published Weekly
$14.00/year

Weekly trade journal covering the entire broadcast industry, listing new equipment, job promotions, stock listings, coming events, summary of daily FCC actions and financial information in the industry.

Also publish:
Broadcast Year Book $13.50, **CATV Source Book** $8.00.

Cable Television Cash Flow Projection

Communication Publishing Corporation
1900 West Yale
Englewood, Colorado 80110

$2.50 each
$2.00 each in orders of ten or more.

Prepared by the Communications Publishing Corporation. It is an excellent custom form for professional presentations, as well as a handy work form. A sample of the major sections follows. A complete set of forms in a compact 8-½ x 11 booklet format, with fold-out sheets for a full eight-year projection, can be obtained by writing to the above address.

FINANCIAL INFORMATION

Date ______________________________

Name of Applicant ______________________________ Phone ______________

Address ______________________________

Location of System ______________________________ Age of System ________

Potential ____________ Present Hook-ups ____________ 5 Yr. Potential ____________

Principals of System (Corporation) ______________________________

President ______________________ Secretary ______________________

Vice-President ______________________ Treasurer ______________________

Owner (If Single Proprietorship) ______________________________

Partners (If Partnership) ______________________________

CREDIT INFORMATION

Personal References (include address)

1. ______________________________
2. ______________________________
3. ______________________________

Company References (include address)

1. ______________________________
2. ______________________________
3. ______________________________

Banking Connection ______________________________

FINANCING INFORMATION

Amount ______________________ Term Requested ______________________

Rate Requested ______________________ Payments ______________________

Lease ______________________ Conditional Sales Contract ______________________

SECURITY OFFERED

Personal Guaranties ________ Yes ________ No Assignment of Pole Agreement ________ Yes ________ No

Assignment of Franchise ________ Yes ________ No Assignment of Capital Stock ________ Yes ________ No

Other ______________________________

Current balance sheet and profit and loss statement for corporation (proprietorship or partnership) attached ________

Current financial statement for each of the principals attached ______________________

BACKGROUND INFORMATION

I. POPULATION HISTORY		II. NUMBER OF HOMES IN CABLE AREA		III. OFF-THE-AIR SIGNAL AVAILABLE			
					Station	Network	Quality
1950	________	To Date	________	(1)	________	________	________
1960	________	Estimate 1970	________	(2)	________	________	________
To Date	________	Estimate 1975	________	(3)	________	________	________
Estimate 1970	________	Estimate 1980	________	(4)	________	________	________
Estimate 1975	________	(if exact count is not available, use		(5)	________	________	________
Estimate 1980	________	3.5 people per home)		(6)	________	________	________

IV. PROPOSED CABLE SIGNAL

	Ch.	City	Network	Dist.	Micro-wave
1.					
2.					
3.					
4.					
5.					
6.					
7.					
8.					
9.					
10.					
11.					
12.					
13.					
14.					
15.					
16.					
17.					
18.					
19.					
20.					

V. REVENUE RATES

Connection Charge ______________________

Monthly Service Charge ______________________

Additional Set Charge ______________________

Commercial Charge ______________________

Other ______________________

VI. FRANCHISE INFORMATION

(1) Date Executed ______________________

(2) Franchise with (a) ______________________

(b) ______________________

(3) Term ______________________

(4) Franchise Payments ______________________

(5) Restrictions (a) ______________________

(b) ______________________

(6) Right to Assign ______________________

VII. POLE USE AGREEMENTS

(1) Telephone Company

Date Executed ______________

Cost per pole ______________

Number of poles ______________

Yearly Cost ______________

Bond Required ______________

Right to Assign ______________

(2) Power Company

Date Executed ______________

Cost per pole ______________

Number of poles ______________

Yearly Cost ______________

Bond Required ______________

Right to Assign ______________

(3) Municipal Poles

Date Executed ______________

Cost per pole ______________

Number of poles ______________

Yearly Cost ______________

Bond Required ______________

Right to Assign ______________

VIII. DESCRIPTION OF TOWN

Number of Street Miles ______________ Homes per mile ______________

Note: Submit Chamber of Commerce data & local bank evaluations: ______________

IX. MAJOR EMPLOYERS

Company	Employees	Company	Employees
(1) ______	______	(6) ______	______
(2) ______	______	(7) ______	______
(3) ______	______	(8) ______	______
(4) ______	______	(9) ______	______
(5) ______	______	(10) ______	______

X. MICROWAVE

Name of Carrier ______

Rate per year per channel hop ______ Number of hops ______ Annual Charge $ ______

CAPITAL OUTLAY ESTIMATE

I. DISTRIBUTION SYSTEM

______ miles of system times $ ______ per mile ______ $ ______

II. HEAD-END & TOWER

(1) Tower-height in ft. ______ times $25 per ft. (If tower exceeds 400 ft. use higher per ft. cost) ______

(2) Tower site preparation – roads, clearing, etc. ______ ______

(3) Antennas – $350 per channel times ______ of channels ______ ______

(4) Head-End Building – (8x8x12) insulated, air conditioned, estimate $2,500.00 ______ ______

(5) Head-End Equipment (If "modulated-demodulated" estimate $1,200 per channel) ______ ______

(If microwave estimate $800 per channel) ______ ______

(6) Weather Display Ch. – $ ______ Basic Pkg; $ ______ camera ______ ______

(7) Antenna Hardware – booms $250/channel X ______ channels ______ ______

(8) Other – winch, loadline, blocks for tower, (estimate $1,000.00) ______ ______

(9) Local Origination – studio equipment and electronics ______ ______

TOTAL $ ______

III. EXTRA CHARGES

(1) Telephone & power pole make ready (estimate based on local negatives.)
Varies considerably over nation ______ ______

(2) Office furniture & Equipment $ ______ desks, $ ______ chairs, $ ______ typewriters, $ ______ other ______

(3) Office Remodeling ______ ______

(4) Tools & Equipment – Sweep gear, field strength meters, etc. ______ ______

(5) Vehicles – $ ______ cars, $ ______ trucks ______ ______

TOTAL $ ______

IV. HOUSE DROP

Estimate $10 per house drop; cable $5.00; tap $3.50; other $1.50. Labor included in expense projection.

Year	New House Drop Customers		Cost	Year	New House Drop Customers		Cost	Year	New House Drop Customers		Cost
(1)	______	X 10	$ ______	(4)	______	X 10	$ ______	(7)	______	X 10	$ ______
(2)	______	X 10	$ ______	(5)	______	X 10	$ ______	(8)	______	X 10	$ ______
(3)	______	X 10	$ ______	(6)	______	X 10	$ ______	Total of 8 years		X 10	$ ______

SUMMARY

______ miles distribution system ______ $ ______ Sub Total $ ______

Head-End & Tower ______ $ ______ House drops $ ______

Other Charges ______ $ ______ Total System Cost $ ______

EXPENSE PROJECTION

8 Years

Line	EXPENSE CATEGORY									
Line	PLANT OPERATING EXPENSE	GUIDELINE DESCRIPTION	Year 1	Year 2	Year 3	Year 4	Year 5	Year 6	Year 7	Year 8
1	Salary – Chief Engineer	12 X monthly salary of $....................								
2	Salary – Ass't Chief Engineer	12 X monthly salary of $....................								
3	Salary – Technicians	No. ...... X 12 times monthly salary								
4	Salary – Temporary Installers	No. ...... X 12 times monthly salary								
5	Pole Rental	50 poles/mile = ...poles X $.......... rate								
6	Tower Site Rental	$........... per month X 12								
7	Vehicle – incl. gas & repairs	$1,000 per vehicle X .... vehicles								
8	Power Head End	$600 per year								
9	Power System	$10 per mile X .... miles per year								
10	Maintenance Parts	$500 per year								
11	Insurance – Fire	$500 per $100,000 of system cost								
12	Insurance – Liability	$300 per $100,000 of system cost								
13	Other									
14	Total Plant Operating Expense									
	OFFICE OPERATING EXPENSE									
15	Salary – General Manager	12 X monthly salary of $....................								
16	Salary – Office Personnel	No. ...... X 12 times monthly salary of $......								
17	Payroll Taxes & Vacation, etc.	10% of gross sal. (incl. plant personnel)								
18	Rent – Office	$........... per month X 12								
19	Telephone	$200/yr. + $100/employee per yr.								
20	Heat, Light, Power – Office	$60 per year per employee								
21	Office supplies incl. mail	50¢ per customer per year								
22	Legal Expense	Estimate								
23	Advertising	Use % of gross revenue; ....%								
24	Travel and Convention Expense	$1,000/yr. depending on size of system								
25	NCTA Dues & Other Subscriptions	$50 per 100 subscribers per year								
26	Property Taxes	% of valuation – ....% of $........... value								
27	Franchise Taxes	Per franchise ....% of gross								
28	Sales Commissions	Estimate								
29	Other Office Expense	Estimate								
30	Bad Debt Expense	Estimate								
31	Accounting Expense	Estimate								
32	Total Office Operating Expense									
	Total Expense (Line 14 plus Line 32)									

Name of Town ______________________________

8 YEAR FORECAST & CASH FLOW PROJECTION

Monthly Rate $ __________ Connection Charge $ ______________

	Year 1	Year 2	Year 3	Year 4	Year 5	Year 6	Year 7	Year 8
Homes Exposed								
Saturation %								
Subscribers Beginning								
Subscribers Added During Year								
Subscribers Ending								
Average for Year (Opening Count + ½ year's addition)								
Monthly Service Income – Ave. sub. count times annual rate per subscriber								
Connection Charges								
Gross Revenue								
Less Expense (See schedule attached)								
Operating Income Before Microwave Expense								
Less Microwave Expense (where applicable)								
Cash Flow – Before Interest, Depreciation & Corporate Income Taxes								

Projected 8 Year Cash Flow, Total $ ____________

NOTES:

Broadcast Management/Engineering

820 Second Avenue
New York, New York 10017
(212) 685-5320

Published Monthly
$15.00/year

A special section in each edition covers systems technical standards, local program organization, promotions and equipment.

Cablecast

Paul Kagan Associates, Inc.
20 East 63rd Street
New York, New York
(212) 826-9725

Published Bimonthly
$100.00/year

Consulting and information services pertaining to investment and CATV finance, contains financial analysis of the broadcast and CATV industries.

Cablecasting

C. S. Tepfer Publishing Company, Inc.
140 Main Street
Ridgefield, Connecticut 06877
(203) 438-3774

Published Bimonthly
$8.00/year

"Cablecasting-Cable TV Engineering: engineering magazine of the cable industry." It is circulated to the technical management personnel of every CATV system in the United States and Canada, TV broadcasting station engineers, microwave relay system and telephone company engineers, consultants, construction contractors and selected officials of government and military agencies.

Radical Software

8 East 12th Street
New York, New York 10003

Published Quarterly
$3.00 per issue

Quarterly journal devoted to cable television video cassette film-making. Each issue includes informative articles on all aspects of video film production, latest equipment, video clubs, organizations and video exchanges.

TV Communications Magazine

1900 West Yale
Englewood, Colorado 80110
(303) 761-3770

Monthly Subscription
$10.00/year, group discounts available

Monthly journal devoted entirely to cable television. Each issue includes informative features on system management, promotions, financing, engineering subjects and monthly summary of CATV news.

selected readings

There is a substantial amount of literature in print on cable technology and telecommunications policy. While very little of this material treats the special needs, interests and problems of minorities, it does provide important background data and information about the industry and the various nonminority interest groups involved. An illustrative cross-section of readily available materials, written primarily for a nontechnical audience is listed here.

Reprints and copies of newspaper and magazine articles can generally be obtained from the publisher. Your local public library can also assist you in obtaining these materials.

General

Browing, Frank. "**Cable TV: Turn On, Tune In, Rip Off,**" *Ramparts*, April 1971.

"**Cable TV,**" *Black Enterprise*, August 1971.

"**Cable TV Leaps into the Big Time,**" *Business Week*, November 1969.

Friendly, Fred. "**Asleep at the Switch of the Wired City,**" *Saturday Review*, October 10, 1970.

Gould, Jack. "**Cable TV: Where the Action Is,**" *New York Times*, Oct. 15, 1967.

Gould, Jack. "**The Point Is, What Will the Cables Carry?**" *New York Times*, June 14, 1970.

Hill, Donald K. "**Communications, Public Interest, and Community Participation,** *Texas Southern Law Review*, December 1970.

Johnson, Nicholas. "**CATV: Promise and Peril,**" *Saturday Review*, Nov. 11, 1967.

"**Leased Channel Concept.**" *TV Communications.*

Manchester, Harland. "**Cable TV – The Hottest Thing in Television,**" *Readers Digest*, June 1969.

National Cable Television Association. **Tune in on the Wonderful Colorful World of Cable TV.** Copies for public distribution have been issued by NCTA since 1969.

Shayon, Robert Lewis. "**Signals: One, Two, Three,**" *Saturday Review*, Dec. 6, 1969

Smith, Ralph Lee. "**Deadlier Than a Western: The Battle Over Cable TV.**" *New York Times Magazine*, May 26, 1968.

Smith, Ralph Lee. "**The Wired Nation.**" *Nation*, May 18, 1970.
TV Communications. Annual Index. Usually in each February edition.
Television Digest. Annual Index. Contains a category listing cable TV articles.

Cable and Education

Drake, Edward E. "**Educators and CATV: A Natural Alliance,** "*TV Communications*, March 1969. Experiences of GenCoE Systems in their work with schools.

"**In Elmira (New York), 21,000 of 25,000 Rate CATV a Necessity,**" *New York Times*, May 4, 1970.

Lewis, P. "**CATV: The Future Channel for Local School Programs,**" *Nation's Schools*, February 1970.

Livingston, Harold. "**Between College and Community: CATV Provides the Perfect Link,**" *TV Communications*, March 1970.

Truby, J. David. "**Educational Television: Growing as CATV Grows,**" *TV Communications*, January 1969.

Community Use of Cable

Bushnell, Don D. and Hallock Hoffman. **The Watts Community Communications Bureau and Training Center: A Proposal for the Mafundi Institute.** March 1970. 49pp., mimeographed.

Keith, Thomas E. "**Orient Your Cablecasting About Your Community.**" In **Originating Programs On Your Cable System: A Collection of**

Outstanding Cablecasting Articles from *TV Communications* (1969)

White, Stephen. **"Toward a Modest Experiment in Cable Television."** *Public Interest*, Summer 1968. Describes plans for wiring Bedford-Stuyvesant area of Brooklyn.

Cable Technology and Channel Capacity

Cooper, Grover C. **"A New Look at Microwave,"** *TV Communications*, November 1970. "Microwave can provide a service-extending, cost cutting adjunct to cable technology."

Fiamengo, Andy. **"Beginning of a New Era: Microelectronics for CATV"** *TV Communications*, September 1970.

"Guerilla Television," *Newsweek*, Dec. 7, 1970.

Rickel, Jack A. **"Video Recording Equipment: What's Best for Your System?** *TV Communications*, November 1970. "The half-inch VTR could prove to be the best on-the-go workhorse for your origination system."

Scott, Jim. **"Bi-Direction: CATV's Future,"** *CATV*, June 8, 1970.

Switzer, I. **"A State-of-the-Art Review: Expanding Cable Capacity,"** *TV Communications.* Three parts. June, July, August, 1970.

Waters, Harry. **"A Bogart to Go, Please."** *New York Times*, Drama Section, Nov. 1, 1970.

Finance/Economics

Bank of New York. **The CATV Industry**. Investment Research Department, November 1970. 35pp.

"Clear Signal?" *Barron's*, Aug. 3, 1970.

Corry, Catherine S. **"Community Antenna Television: An Economic Survey."** Washington: Legislative Reference Service, Library of Congress, Sept. 27, 1965. Mimeographed. 52 pp.

Drexel Harriman Ripley (now Drexel Firestone). **An Industry Report on Community Antenna Television (CATV)**. Oct. 15, 1968.

Dreyfus Corporation. **"A Report on CATV."** No date, mimeographed. 10 pp.

Legg & Company. **CATV: A 'Cool' Medium Turned Hot.** Legg & Co. Research Department, August 1969.

Little, Stuart W. "**CATV: The Big Growth Industry in Broadcasting.**" *Saturday Review*, Oct. 12, 1968.

Future Uses of Cable

Belt, Forest H. "**Television: 20 Years from Now.**" *Electronic World*, January,1970.

"**The Future of Cable TV.**" *The Magazine of Wall Street*, July 5, 1969.

Gould, Jack. "**If Cable TV Ever Goes Where It Could Go,**" *New York Times*, June 15, 1969.

Kern, Edward. "**A Good Revolution Goes on Sale,**" *Life*, Oct. 16, 1970.

"**Video Cassettes Offer Potential for Future But Problems for Now,**" *Wall Street Journal*, Dec. 16, 1970.

Local Origination

Keith, Thomas E. "**Orient Your Cablecasting About Your Community.**" In "Originating Programs On Your Cable System: A Collection of Outstanding Cablecasting Articles" from *TV Communications*. (1969)

"**Local Origination: The Hard Facts,**" *TV Communications*, June 1970.

Mooney, John W.P. "**Local Community News – The Backbone of Origination,**" *TV Communications*, December 1969.

Originating Programs on Your Cable System. "A collection of outstanding cablecasting articles" from *TV Communications.* (1969)

"**Successful Cablecasting** – A Summary of Activities: Capsule Reports on ten selected local origination operations," *TV Communications*, October 1969.

Regulations

Barnett, Harold J. and Edward Greenberg. "**Regulating CATV Systems.**" *Law and Contemporary Problems*, No. 34, 1969.

Christiansen, Gary L. "**Should CATV Be a Utility?**" *Broadcast Management/Engineering*, December 1969.

Dietsch, Robert W. "**Public Utilities: A Path through the Maze,**" *Nation*, Aug. 17, 1970.

Harvard Law Review. "**Notes: The Wire Mire: The FCC and CATV,**" Vol. 79 (1965), pp. 366-390.

Johnson, Leland. **The Future of Cable Television: Some Problems of Federal Regulation**. Rand Corporation Memorandum RM-6199-FF. January 1970. 87 pp.

NCTA. **The CATV Industry and Regulation.** Washington (1970). Booklet, unpaged. Sets forth the cable industry's position.

Oppenheim, Jerrold. **Model Code of Regulations Cable Television – Broadband Communications,** ACLU, Illinois Division, June 1971.

Rivikin, Steven R. "**Jurisdiction to Regulate Cable Television: Issues in the Scope of Federal, State and Municipal Powers.** Sloan Commission, 1971.